SpringerBriefs in Service Science

SpringerBriefs present concise summaries of cutting-edge research and practical applications across a wide spectrum of fields. Featuring compact volumes of 50 to 125 pages, the series covers a range of content from professional to academic.

Typical publications can be:

A timely report of state-of-the art methods

A bridge between new research results, as published in journal articles

A snapshot of a hot or emerging topic

An in-depth case study

A presentation of core concepts that students must understand in order to make independent contributions

SpringerBriefs are characterized by fast, global electronic dissemination, standard publishing contracts, standardized manuscript preparation and formatting guidelines, and expedited production schedules.

The rapidly growing fields of Big Data, AI and Machine Learning, together with emerging analytic theories and technologies, have allowed us to gain comprehensive insights into both social and transactional interactions in service value co-creation processes. The series SpringerBriefs in Service Science is devoted to publications that offer new perspectives on service research by showcasing service transformations across various sectors of the digital economy. The research findings presented will help service organizations address their service challenges in tomorrow's service-oriented economy.

This series has been accepted by Scopus.

Xuan Li · Fei-Yue Wang

Scenarios Engineering for Services Science

Towards Trustworthy and Reliable Services Systems with New AI

Springer

Xuan Li
Beijing Institute of Technology
Zhuhai, China

Fei-Yue Wang
Institute of Automation
Chinese Academy of Sciences
Beijing, China

ISSN 2731-3743 ISSN 2731-3751 (electronic)
SpringerBriefs in Service Science
ISBN 978-3-032-24151-1 ISBN 978-3-032-24152-8 (eBook)
https://doi.org/10.1007/978-3-032-24152-8

This Springer imprint is published by the registered company Springer Nature Switzerland AG
The registered company address is: Gewerbestrasse 11, 6330 Cham, Switzerland

Preface

The evolution of services science has been fundamentally reshaped by the convergence of Artificial Intelligence (AI), deep learning, and Large Language Models (LLMs). Services science-once centered on human-driven design and empirical business intuition-now faces a new era in which intelligent agents, digital ecosystems, and autonomous intelligence are becoming indispensable actors in value co-creation. Yet, this transformation is not merely technological: it demands a new philosophy for how service systems are conceived, verified, and governed.

This brief presents Scenarios Engineering for Services Science (SE4SS) as a paradigmatic framework to safeguard the development of future service systems, including intelligent finance, intelligent customer services, intelligent hotels, intelligent cities, intelligent logistics, intelligent healthcare, and intelligent vehicles. Rather than relying on brittle feature selection or intuition-based workflow design, SE4SS constructs service intelligence through carefully structured scenarios. These representations capture the evolving semantics of service activities-real or synthetic, physical or digital, individual or societal-and couple them with systematic mechanisms of Intelligence & Index, Calibration & Certification, and Verification & Validation. Based on ACP (Artificial societies, Computational experiments, and Parallel execution), SE4SS aims to liberate services science from isolated models and opaque heuristics, transforming it into calibrated, verifiable and reliables.

Key highlights include:

Scenarios Engineering for Services Science

Services science is ultimately relational: customers, providers, cyber infrastructures, and social constraints co-evolve within bounded spaces and times. Scenarios are the connective tissue of this ecosystem. They operate as the semantic substrate upon which new AI components-foundation models, autonomous agents, generative simulators, and decentralized intelligence-can be trained, evaluated, and deployed. In addition, SE4SS asks a deeper question: "Under which scenarios can intelligent service act efficiently, effectively, safely, sustainably, and reliably?" In this regard, scenarios engineering do for new AI what experimental frameworks did for traditional engineering-they turn uncertain complexity into controllable, measurable, and certifiable intelligence.

Parallel Service Systems by Parallel Intelligence

Traditional AI has excelled in prediction but often failed under rare, extreme, or poorly calibrated service conditions. Inspired by Parallel Intelligence, SE4SS introduces a service-centric methodology. Through this integration, SE4SS dissolves the boundaries between laboratory models, operational practices, and regulatory compliance. It shifts services from reactive management to proactive, scenario-aware orchestration, enabling autono To ensure intelligent service, we advocate guiding service systems toward a "6S" paradigm: service systems that are Safe in the physical world, Secure in the cyberspace, Sustainable in the ecological development, Sensitive to privacy, individual rights, and responsible resource utilization, capable of providing Service to All, and fostering the Smartness of All.

Future Service Systems with New AI

We foresee that future service systems, under the paradigm of parallel worlds, will operate in three modes. For the vast majority of time, service systems will exist in Autonomous Modes (AM), governed by swarms of digital agents that coordinate more than 80% of service operations. Parallel Modes (PM) will be employed for concrete execution, where autonomous service robots carry out critical tasks, with their presence limited to less than 15%, acting as the system's mechanical guardians. Expert/Emergency Mode (EM) is activated only under extreme events. In such cases, service scientists perform critical tasks on-site, supported by digital agents and autonomous service robots, and their involvement accounts for no more than 5% of all interactions.

While this brief aspires to advance the theoretical, methodological, and practical foundations of SE4SS, we remain acutely aware that this work represents only an initial step toward a more systematic understanding of service intelligence. We sincerely welcome comments, corrections, and suggestions from colleagues, students, and readers from all related communities. Any valuable insights-whether theoretical refinements, empirical evidence, or conceptual disagreements-will serve as a meaningful contribution to the continued development of this research.

Zhuhai, China Xuan Li
Beijing, China Fei-Yue Wang
November 2025

Competing Interests The authors have no competing interests to declare that are relevant to the content of this manuscript.

Contents

Chapter 1
From Scenarios Engineering to Reliable Service Systems Through ACP

1.1 Introduction

Services science (SS) [1, 2] is an interdisciplinary field that studies the design, implementation, management, and improvement of service systems. It integrates knowledge from multiple disciplines including computer science, management science, engineering, and social sciences to understand and optimize service-related activities. At its core, services science focuses on value co-creation between providers and customers. It examines how people, technologies, organizations, and shared information work together to deliver services that satisfy human and business needs.

However, services science has traditionally been human centered, relying heavily on human labor for service design, delivery, and interaction. While this approach enables flexibility and empathy, it also introduces significant limitations, including inconsistent quality, high labor costs, limited scalability, and vulnerability to human error and fatigue. In recent years, the rapid development of artificial intelligence (AI) has provided a transformative solution for various fields by enabling data-driven decision-making, real-time responsiveness, and scalable personalization. The combination of services science and AI is transforming the nature of modern service systems across industries. AI contributes to the development of services science in several ways, as shown in Fig. 1.1. It enables the automation of routine and repetitive tasks, which reduces operational costs and improves efficiency. For example, AI-powered chatbots can handle customer inquiries 24/7, freeing up human employees for more complex and value-added interactions. Besides, AI has significantly enhanced service quality and scalability of intelligent vehicle through real-time perception, decision-making and control. In this context, new AI components provide transformative pathways to address long-standing challenges in services science by enabling more adaptive, intelligent, and scalable infrastructures. Nevertheless, current AI methodologies still exhibit notable limitations, including insufficient interpretability, limited robustness in sparse or atypical scenarios, and strong dependencies on high-quality data and computational resources. These constraints often

X. Li and F.-Y. Wang, *Scenarios Engineering for Services Science*,
SpringerBriefs in Service Science, https://doi.org/10.1007/978-3-032-24152-8_1

Fig. 1.1 AI contributes to services science in many ways. Top: A 24/7 chatbot. Bottom: A self-driving car service named "RoboTaxi"

impede the reliable deployment of AI techniques in complex, real-world service environments.

In general, this book will introduce new developments brought by AI to the field of services science—specifically in intelligent finance, intelligent customer services, intelligent hotels, intelligent cities, intelligent logistics, intelligent healthcare, and intelligent vehicles in the following chapters. In this chapter, we first discuss the current challenges faced by services science. Then, from the new perspective of ACP (Artificial societies, Computational experiments, and Parallel execution), we outline the paradigm shift brought about by new AI in services science research. We also briefly describe parallel intelligence in supporting artificial modeling, computational experiments, and parallel execution with new AI. To this end, we introduce a general framework, named SE4SS (Scenarios Engineering for Services Science), to support the systematic analysis and development of service systems. The SE4SS framework enables the training, evaluation, and comparative assessment of algorithms that integrate AI, scenario engineering, and service-oriented system design.

1.2 Parallel Intelligence, Artificial Intelligence, and Services Science

1.2.1 ACP: The Engine Behind Parallel Intelligence

Parallel Intelligence [3], as a novel computational paradigm for modeling, analyzing, and managing complex systems, is fundamentally built upon the ACP method—an acronym for Artificial societies (A), Computational experiments (C), and Parallel execution (P). This method [4] enables researchers and practitioners to construct artificial societies mirroring real-world dynamics, conduct virtual experiments for predictive analysis, and implement parallel control strategies that bridge simulated and physical environments. By iteratively refining decisions through simulation and interaction, the ACP method provides a closed-loop feedback mechanism crucial for adaptive and intelligent service systems.

1.2.2 Parallel Intelligence for New Artificial Intelligence

Traditional artificial intelligence (AI) systems primarily focus on perception, learning, and optimization, often under static or narrowly defined conditions. Parallel learning [5] is a new AI approach that combines descriptiveness, predictability, and prescriptiveness. Together with parallel intelligence, it supports data-driven modeling, simulation, and decision-making in complex service environments. Descriptive learning builds virtual service systems, predictive learning generates scenario data, and prescriptive learning explores decision strategies. This unified framework supports intelligent, transparent, and resilient cross-domain new AI designs, such as in the fields of intelligent transportation, intelligent healthcare, and intelligent cities.

1.2.3 New Artificial Intelligence for New Service Systems

Within services science, AI transforms traditional models by supporting personalized, adaptive, and low-human-involvement service delivery. It improves resource allocation, enhances user experience, and facilitates real-time coordination among stakeholders. AI contributes to personalized service delivery, predictive customer engagement, intelligent resource allocation, and real-time decision support, all of which are essential in modern service ecosystems such as smart healthcare, intelligent finance, and digital governance. Besides, AI enables safe navigation in mobility services, optimizes traffic and energy use in urban systems, and advances diagnostics in healthcare. Moreover, when integrated with the ACP method, AI can transcend reactive automation and achieve proactive, goal-oriented service innovation. This synergy empowers service systems to evolve from human-centered coordination to machine-assisted co-creation, fostering more efficient, adaptive, and resilient service processes. However, the integration of AI into services science brings challenges in terms of transparency, security and ethics. High-impact applications demand explainable decisions, resilience against cyber threats, and safeguards for privacy and fairness.

1.3 Scenarios Engineering for Trustworthy AI

1.3.1 Untrustworthy AI Caused by Feature Engineering

Feature engineering refers to the process of selecting, transforming, and creating variables—known as features that can be used by machine learning or AI models to improve their performance. It is often considered the "art" of deriving meaningful representations from raw data in a way that aligns with the learning objectives of the model, especially when employing deep-learning approaches. Through careful transformation of input data into more informative formats, feature engineering helps uncover patterns that are crucial for understanding service behaviors, customer preferences, and operational dynamics. With the rapid advancement of AI, a growing number of state-of-the-art models have been developed that rely heavily on both sophisticated network architectures and effective feature engineering. When done well, it not only increases the value of existing data but also significantly boosts the performance of AI models deployed in service environments. However, poor feature [6–8] choices can lead to underperforming models or necessitate complex architectures to compensate for the lack of informative input representations, as illustrated in Fig. 1.2. Another challenge lies in the increasing reliance on black-box models, such as deep neural networks, resulting from feature engineering. These models excel at processing large-scale, high-dimensional feature spaces—often derived from unstructured or semi-structured data such as text, images, or audio—and can deliver high accuracy in predictive tasks. Besides, this performance often comes at the cost of explainability, making it difficult for service designers, operators, and decision-makers to interpret how or why a specific outcome was generated. This lack of transparency poses substantial risks in service domains where accountability, fairness, and regulatory compliance are critical—such as intelligent vehicle, healthcare, education, and public administration.

1.3.2 Why Scenarios Engineering Needed

Interpretability has always been regarded as a fundamental requirement for trustworthy AI. It provides visibility into the internal workings of AI systems and supports transparency in data-driven decision-making. However, trust extends beyond mere

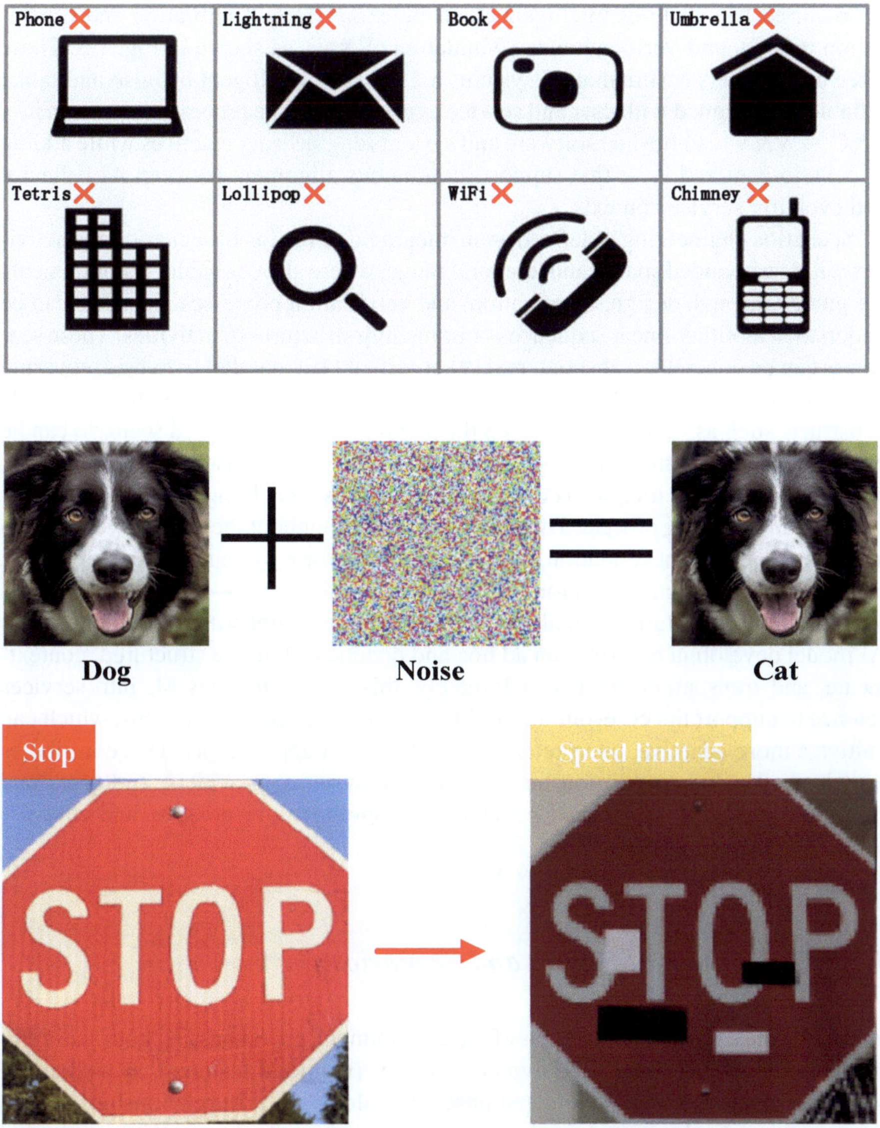

Fig. 1.2 Examples of failure scenarios in artificial intelligence

interpretability—it also depends on the controllability of the learning process, particularly the quality and relevance of input features or data. This is especially critical in complex service systems where AI must interact with diverse users, dynamic environments, and multi-layered infrastructures. SE emerges as a systematic approach to address these needs, particularly within the context of services science. It introduces a structured life cycle of Intelligence and Indexing (I&I), Calibration and Certification (C&C), and Verification and Validation (V&V), as shown in Fig. 1.3. These three components ensure that AI systems are not only intelligent but also auditable, reliable, and aligned with user and service expectations. The process mirrors (I&I → C&C → V&V) established software and systems engineering practices while adding a scenario-centered layer that supports continuous alignment between AI behavior and evolving service contexts.

Scenarios engineering is defined as an integrated reflection of scenarios and activities within a bounded spatial and temporal range, where all actionable AI components are guided through design, certification, and verification processes. Scenario can be understood as either linear sequences or branching structures of activities. These scenarios can be concrete or abstract, real [9] or virtual [10], parallel or hybrid, covering a spectrum from physical environments (e.g., atoms and molecules) to large-scale constructs such as smart cities or even the metaverse. Importantly, a scenario can be expressed in various modalities—text, natural language, databases, video, animation, historical records, or imagined constructs like dreams—enabling a holistic representation of any event or process. By embedding SE throughout the AI life cycle—such as during requirement validation, operational goal setting, system testing, third-party certification, and final evaluation of design and performance—it promotes AI systems. Moreover, SE fundamentally transforms feature engineering—a critical step in AI model development—from an ad hoc and opaque task into a structured, context-aware, and transparent practice. Ultimately, this brief integrates SE into services science to support the co-evolution of AI and intelligent service systems, which can cultivate more adaptive, interpretable and reliable AI applications. This contributes to enhanced value co-creation between service systems, providers, and users and lays the foundation for the next generation of intelligent, responsive, and ethically aligned service ecosystems.

1.3.3 Definition of Scenes and Scenarios

Scenes: Scenes represent snapshots of the environment, encompassing both static elements (structured features) and dynamic entities (temporal features), as well as the self-representations of all actors and observers, along with the relationships among these entities. In the virtual world, scenes can be described comprehensively and objectively—serving as the ground truth. However, in the real world, scene representations are inherently incomplete, potentially inaccurate, and uncertain, as they are constructed from the perspectives of one or more observers—thus forming subjective scenes.

Scenarios: Scenarios describe the temporal evolution across a sequence of scenes, beginning with initial scenes. It captures the progression over time by specifying actions, events, goals, and values that shape the development of the scenario. Unlike a scene, which represents a single moment in time, a scenario spans a defined temporal duration (temporal features).

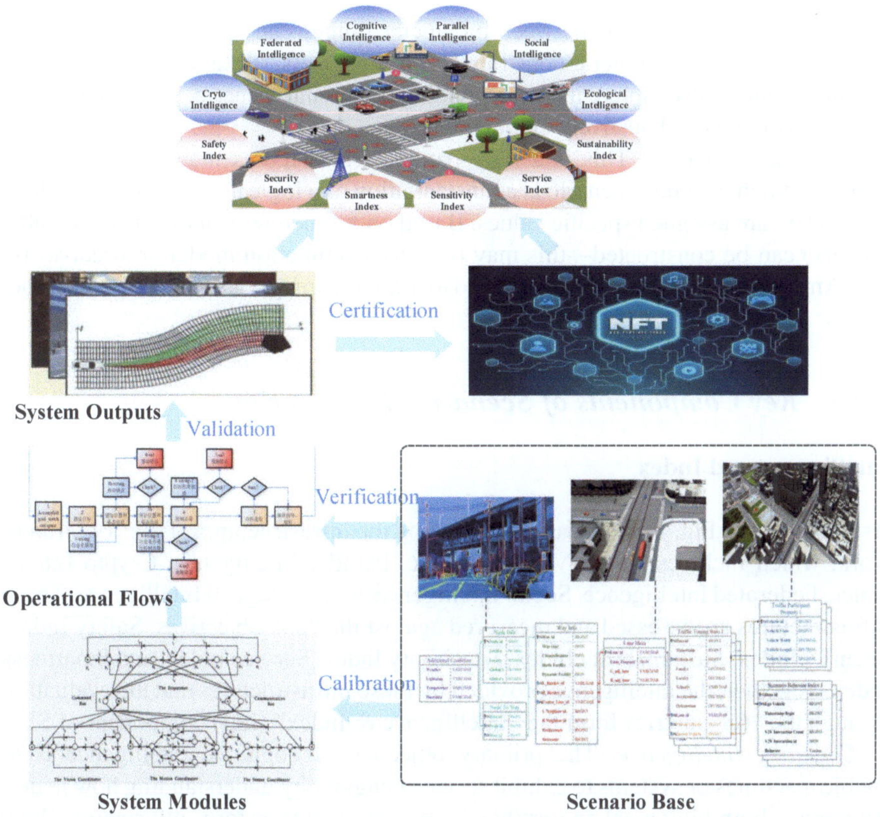

Fig. 1.3 Overall framework of scenarios engineering comprised of Intelligence and Index (I&I), Calibration and Certification (C&C), Verification and Validation (V&V)

1.3.4 Classification of Scenarios

Menzel et al. classified [11] scenario representations into three levels of abstraction: functional scenario, logical scenario, and concrete scenario. The functional and logical scenarios describe scenario spaces at different abstraction levels, whereas the concrete scenario represents a specific instance of a scenario.

Functional Scenarios: This level represents the scenario space at a semantic level, typically using linguistic descriptions. According to [11], "the vocabulary used for the description of functional scenarios is specific to the use case and the domain, and may exhibit varying levels of detail."

Logical Scenarios: This level represents the scenario space in terms of a state space, where each parameter corresponds to an influential factor and is defined by a range of values. These parameter ranges can optionally be associated with probability distributions. A logical scenario includes a formal notation of the scenario space.

Furthermore, "the relations between parameter ranges can optionally be specified using correlations or numeric constraints." In this briefs, we assume that a logical scenario cannot fully capture its corresponding functional scenario, as not all relevant parameters can be exhaustively enumerated.

Concrete Scenarios: This is a fully parameterized representation of a specific scenario. Each concrete scenario is an instantiation of a logical scenario, in which all parameters are assigned specific values. Based on a concrete scenario, an executable scenario can be constructed—this may refer to a simulation model or a real-world test. An executable scene may correspond to a camera image or a LiDAR point cloud.

1.3.5 Key Components of Scenarios Engineering

Intelligence and Index

As illustrated in Fig. 1.4, SE is underpinned by the infrastructures of the "6I" framework, which includes Cognitive Intelligence, Parallel Intelligence, Crypto Intelligence, Federated Intelligence, Social Intelligence, and Ecological Intelligence. These infrastructures are assessed and measured against the "6S" objectives: Safety Index, Security Index, Sustainability Index, Sensitivity Index, Service Index, and Smartness Index. Together, the intelligence provided by these infrastructures and the evaluation through the "6S" metrics form the "Intelligence & Index" (I&I) component of SE.

Cognitive Intelligence: The primary objective of cognitive intelligence is to enable machines or systems to exhibit service behavior by understanding how neural networks—both biological and artificial—process and transform information, how intelligible knowledge is formed and represented, and how decisions are made. In the context of human-machine hybrid systems, this analysis typically focuses on mental faculties such as language, perception, memory, attention, reasoning, and emotion. Cognitive intelligence spans multiple levels of analysis, from learning and decision-making to neural circuits and brain organization. Ultimately, it seeks to unravel the nature of intelligence, with the aim of enhancing both our understanding of the human mind and the design of service machines and systems.

Parallel Intelligence: It provides an efficient framework for transforming small datasets into large datasets and refining large datasets into deep intelligence for specific tasks [12]. It operates in three modes: (1) artificial societies model real-world behaviors to facilitate training and learning; (2) computational experiments are used to validate and predict decision-making processes in artificial societies for testing and validation; and (3) real-time decision-making and parallel execution occur in both real and artificial societies, enabling control and management.

Crypto Intelligence: It focuses on generating reliable, real-time decision-making in trustless environments while ensuring privacy preservation and data security [13]. Built on blockchain technology and structured around Decentralized Autonomous Organizations (DAOs), crypto intelligence leverages federated data as the foundation

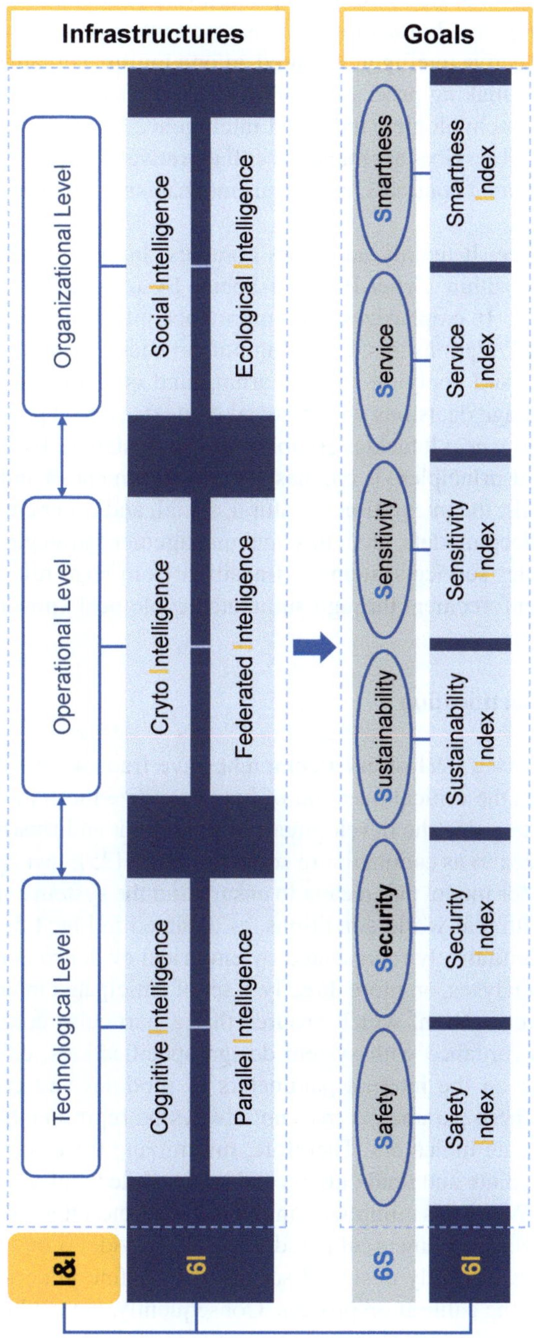

Fig. 1.4 Infrastructures and indexes of scenarios engineering

for decision-making, smart contracts as the decision-making mechanism, and non-fungible tokens (NFTs) as decision-making incentives [14–16].

Federated Intelligence: It transforms the isolated intelligence of individual organizations into a collective intelligence shared among multiple organizations, thereby optimizing decision-making processes within service systems. Driven by big data, AI, and blockchain technologies, federated intelligence fosters a federated ecosystem in which data and services are managed collaboratively [17, 18]. This ecosystem is governed by federated contracts, consensus mechanisms, incentives, and security protocols [19].

Social Intelligence: It mainly addresses complex, human-in-the-loop decision-making challenges within Cyber-Physical-Social Systems (CPSS), where social systems are central. It emphasizes the importance of modeling and analyzing social behaviors and relationships, capturing human social dynamics, creating artificial social agents, and applying social learning and cognition to generate social knowledge and manage decisions within social contexts [20, 21].

Ecological Intelligence: It tackles complex service system tasks by applying ecological thinking and principles. It emphasizes the alignment of individual, group, and community goals, the integration of natural, social, and cyber environments, and the exploration of deep models, analytics, and management strategies for ecological systems. This enables service systems to transition from legal rule enforcement to ethical ecological enforcement through automated ecological knowledge [22].

Calibration and Certification

As previously discussed, I&I offers a comprehensive framework for evaluating AI systems. This raises the critical question of how to utilize these multi-dimensional evaluation indices to guide the development of intelligent and trustworthy AI. Calibration, also referred to as estimation in some literature [25], involves determining appropriate values for model parameters to ensure that the system's internal dynamics closely align with real-world conditions, as depicted in Fig. 1.5. Here, the "real world" is often quantitatively represented by empirical evidence derived from case studies, inductive analyses, or, more directly, a set of anticipated intermediate states.

In contrast to verification, which ensures that a particular component or process operates in accordance with system design specifications, calibration delves deeper by focusing on the internal parameters of modules and components. For complex systems, these parameters may not always be representable in an analytical form through state indicators. Therefore, minimizing the discrepancy between the model's intermediate and final outputs and the expected real-world counterparts becomes essential. While this minimization can involve time-intensive trial-and-error approaches, particularly in emergent paradigms, SE provides a more efficient alternative. SE leverages carefully designed scenarios to define intermediate and final states, streamlining the calibration process. Consequently, the calibration of SE can be formally represented as follows:

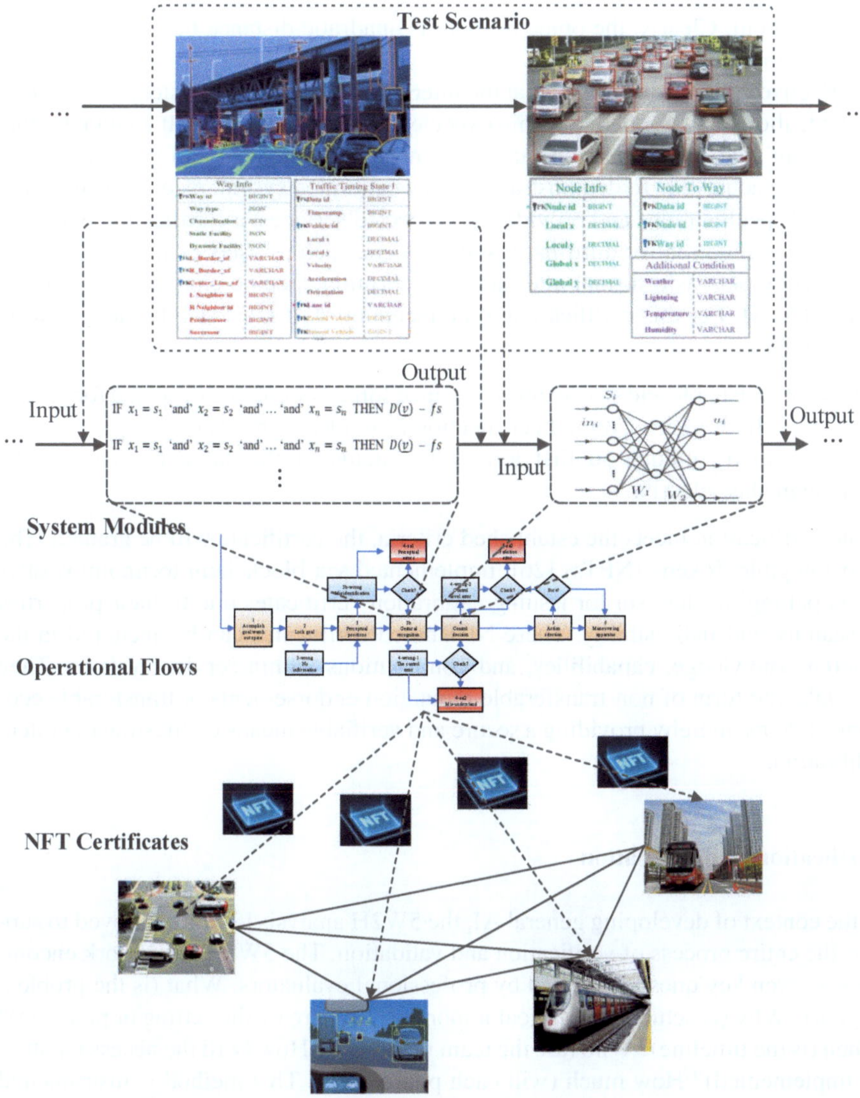

Fig. 1.5 Calibration and certification of scenarios engineering

$$\theta^* = \underset{\theta}{\operatorname{argmin}} \sum_{t=1}^{K} [y(t,\theta) - \hat{y}(t)]^T \cdot V \cdot [y(t,\theta) - \hat{y}(t)] \tag{1.1}$$

where $y(t,\theta) = \left[\, y_1(t,\theta) \cdots y_n(t,\theta) \,\right]^T$ and | $\hat{y}(t) = \left[\, \hat{y}_1(t) \cdots \hat{y}_n(t) \,\right]^T$ are the intermediate outputs of modules or components for test systems and elaborate scenarios in time step t. The positive symmetric matrix V represents the importance of each output. Clearly, the objective here is quadratic distance but it can be easily extended to other metrics as well.

Effective calibration ensures that the internal parameters of a system are properly aligned, allowing it to operate within expected performance levels. It guarantees that the system's designed functions are implemented as intended and that the desired outputs are achieved. To demonstrate that the calibrated system meets the necessary technical specifications and provides the expected services, a certification is issued, serving as validation for third-party stakeholders, such as users or integrators. This certification acts as a formal attestation, confirming specific characteristics of the actionable AI. For this certification to be meaningful, it must satisfy the following requirements:

- It is not perishable and easy to preserve since some systems are designed for several decades. In this sense, digital certification is an ideal candidate.
- It is globally unique so that any two systems of the same class would be distinguished clearly.

If the certification meets the established criteria, the certificate will be granted. The Non-Fungible Tokens (NFTs) [26], implemented via blockchain technology, offer a compelling mechanism for issuing calibration certificates due to their properties of scarcity and indivisibility. These NFT-based certificates can be anchored in the inherent knowledge, capabilities, and contributions within service systems. They may take the form of non-transferable reputation endorsements or transferable economic tokens, thereby providing a secure and verifiable means of attesting to system calibration.

Verification and Validation

In the context of developing general AI, the 5W2H analysis [27] is employed to support the entire process of verification and validation. The 5W2H framework encompasses seven key questions posed by professional evaluators: What (is the problem or goal)? Why (is achieving this goal important)? Where (is the setting or platform)? When (is the timeline)? Who (are the team members)? How (will the necessary steps be implemented)? How much (will each phase cost)? This method is instrumental in gathering relevant activities, thereby facilitating a more effective construction of SE.

In the SE process (see Fig. 1.6), verification assesses the performance of AI, equipment, and engineering with respect to service. It encompasses a range of activities aimed at ensuring the production of high-quality AI, such as design analysis, program testing, system inspections, and specification reviews. Verification is inherently objective-provided that all processes and documentation are expressed with sufficient precision; subjective judgment is unnecessary. During the design and pre-development phases, verification procedures may involve modeling or simulating a system or service and analyzing the results. In later phases, regular and critical testing ensures that the AI continues to meet practical specifications and requirements.

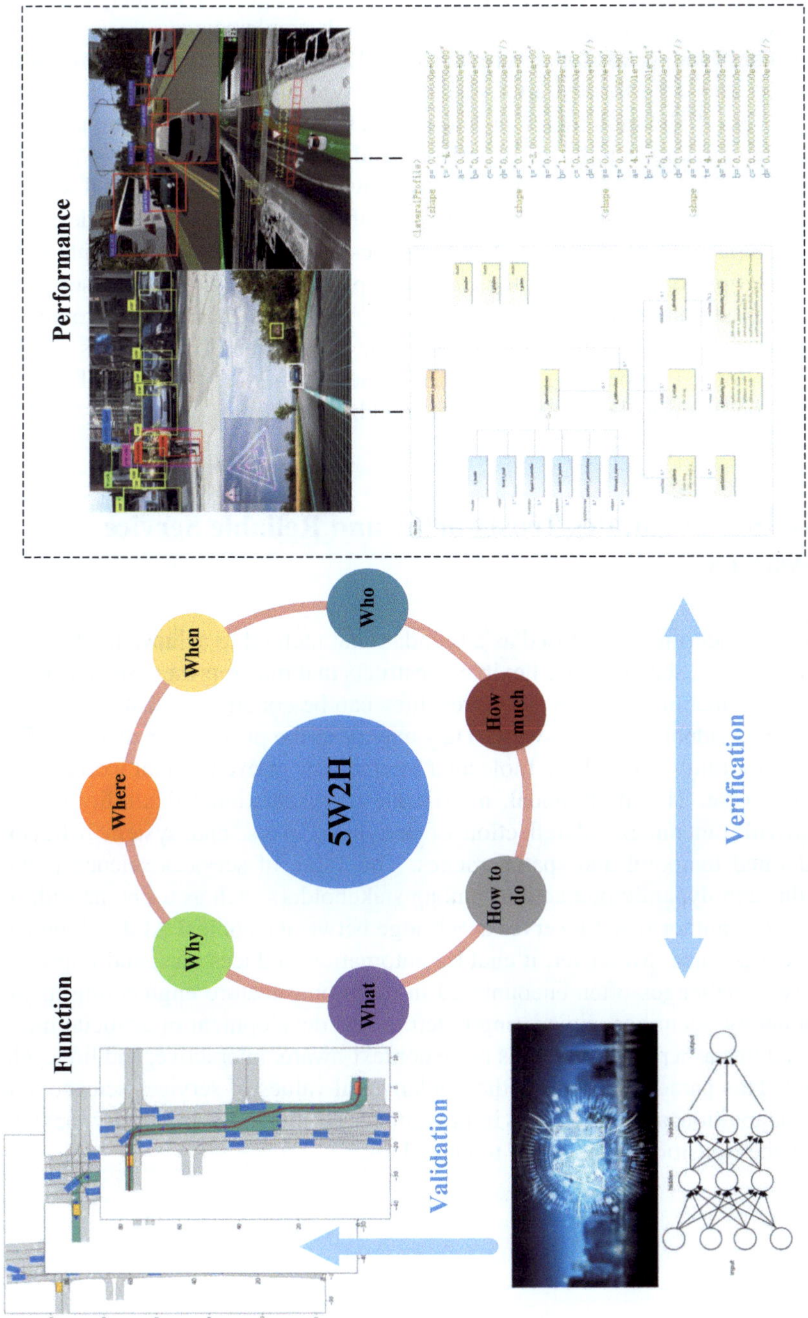

Fig. 1.6 Verification and validation of scenarios engineering

Verification is a continuous process, applied throughout the AI product life cycle, often described by the question: "Are you building the AI right?" This query confirms whether the system's performance is being accurately implemented.

On the other hand, validation ensures that the AI's functionality meets user-defined requirements and is performed after verification, typically at the end of the development process. Validation identifies and resolves abnormal data, functionality, or services, thereby improving user experience and creating feedback loops for the enhancement of trustworthy AI. It may also involve predicting potential system faults or gaps. Validation tests include a variety of functional assessments designed to ensure compatibility with existing functionalities. User-defined requirements serve as the foundation for determining the validation process of environments, services, or systems. In simple terms, we conduct in daily operations—such as environment, function, service, and system testing—constitutes validation. This process can be expressed as: "Are you building the right AI?" Here, "building the right AI" refers to meeting the functional requirements specified by users.

1.4 SE4SS: Facilitates Trustworthy and Reliable Service Systems

Scenarios engineering is proposed as a foundational method to achieve trustworthy AI. In this context, scenarios are flexible constructs that may represent sequences of activities or branching decision structures; they can be concrete or abstract, real or simulated, individual or system-wide. They may describe phenomena at any scale–from urban mobility networks to molecular interactions or even virtual worlds such as the metaverse. This multimodal, multi-scale representational flexibility allows SE to provide an integrated reflection of activities, users, and system behavior within defined temporal and spatial ranges. The value of services science is co-created through dynamic interactions among stakeholders such as users, providers, and regulatory authorities. SE serves as a bridge between technical AI development and services science. Moreover, it enables automation, reduces bias, and improves scalability—challenges often encountered in traditional feature engineering methods. Ultimately, scenarios engineering redefines the development of artificial intelligence systems as a paradigm that is more context-aware, interactive, and life cycle conscious. This paradigm supports the fundamental values of services science: co-creation, adaptation, and continuous improvement. As illustrated in Fig. 1.7, new AI facilitates the transition from old IT to new IT.

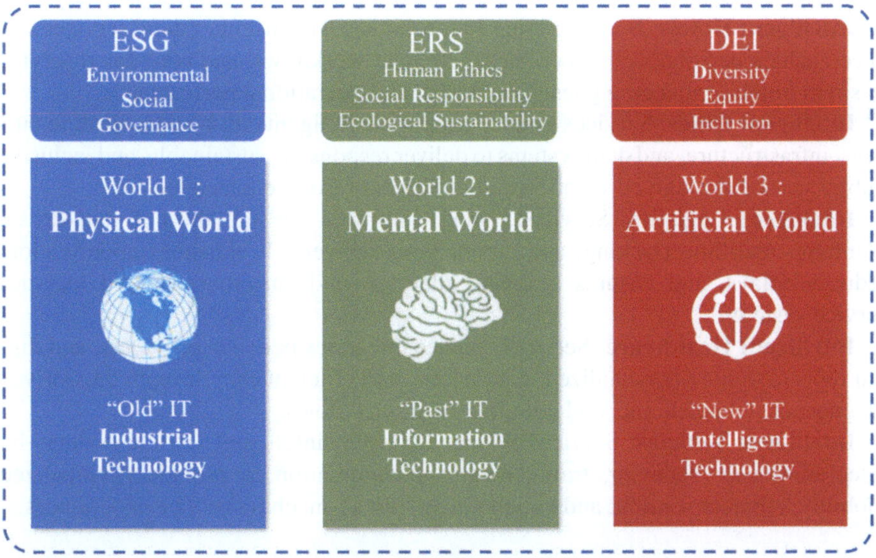

Fig. 1.7 Paradigm shifts from old industry to new industry driven by new AI

Fig. 1.8 Application scenarios in services science

1.4.1 Application Scenarios in Service Systems

Services science encompasses a wide array of complex application scenarios where human needs, business processes, and technological capabilities intersect. As shown in Fig. 1.8, the representative application scenarios include:

Intelligent Finance: Intelligent finance applies services science to deliver personalized, automated, and secure service through robo-advisors, risk assessment, and dynamic, user-centric financial ecosystems.

Intelligent Customer Services: Services science empowers intelligent customer service with chatbots, sentiment analysis, and adaptive knowledge systems to deliver fast, personalized, and emotionally aware interactions across digital platforms.

Intelligent Hotels: By integrating IoT, robotics, and customer data, services science enables personalized experiences, efficient workflows, and automated service in smart hotels, enhancing guest satisfaction and operational performance.

Intelligent Cities: Services science supports intelligent city by coordinating citizens, infrastructure, and data systems to deliver responsive, sustainable, and inclusive public service through co-creation and intelligent resource management.

Intelligent Logistics: Services science improves logistics through data-driven planning, real-time tracking, and autonomous delivery. It enhances coordination, reduces delays, and ensures scalable, adaptive, and customer-centered logistics service systems.

Intelligent Healthcare: Services science integrates patients, providers, and digital tools to deliver personalized, continuous, and efficient care through data-driven diagnostics, telemedicine, and coordinated service flows.

Intelligent Vehicles: Services science enhances intelligent vehicle systems by integrating user behavior, traffic data, and automation to provide personalized mobility, efficient routing, and collaborative human-machine service experiences.

1.4.2 Challenges in Services Science

Despite rapid progress in digital technologies, services science continues to encounter critical challenges rooted in its complex, human-centered, and cross-domain nature. These challenges can be categorized as follows:

Human-Centric Variability: Service quality is often influenced by individual perceptions, emotions, and contextual interpretations, resulting in inconsistent and subjective outcomes that are difficult to standardize or reproduce.

Low Automation and Limited Scalability: Many service processes remain labor-intensive and manually executed, making them hard to automate or scale efficiently across large populations or dynamic contexts.

Fragmented Data: Services frequently operate across disconnected platforms and institutions, leading to incomplete, inconsistent, and non-integrated data, which hinders comprehensive analysis and intelligent decision-making.

Lack of Quantitative Evaluation Standards: Evaluating service performance, especially in emotional, experiential, or personalized domains, lacks universally accepted metrics, making it hard to benchmark, optimize, or compare outcomes reliably.

Difficulty in Personalizing Services at Scale: Delivering real-time, individualized services at scale remains difficult due to diverse user needs, limited contextual awareness, and the absence of adaptive service mechanisms.

1.4.3 What New in the New AI?

"New AI" is an emerging paradigm characterized by autonomous, largeâŁ'scale intelligent systems driven by knowledge automation and parallel intelligence, deeply integrated with real-world application scenarios. Research emphasizes that the core of AI development lies in domain-specific use cases, not just technical models. We advocate embedding AI across disciplines—from humanities to sciences—and fostering adaptive curricula and platforms that link academic learning to industry deployment. In this view, new AI transcends narrow algorithmic thinking and becomes a multidisciplinary, humanâŁ'centered ecosystem where AI agents collaborate in real or virtual environments, enabling dynamic knowledge creation, decision automation, and intelligent service tied to real industry and societal needs. Based on scenarios engineering theory, elevating services science into service system requires the following new AI components.

Complex Scenario Modeling: The design and simulation of real-world environments, enabling AI systems to operate and adapt within diverse and dynamic contexts.

Foundation Models: Large-scale pretrained models (e.g., Large Language Models) that serve as general-purpose engines, capable of transferring knowledge across tasks and domains.

Specific Application Scenarios: Real-world domains where new AI is embedded, reflecting the view that value is realized not through algorithms alone, but through integration with meaningful use cases.

High-Quality Datasets: Carefully curated and augmented datasets that provide the empirical foundation for robust, transferable, and scalable AI learning.

Advanced Model Evaluation Metrics: Sophisticated, context-aware criteria for assessing model performance beyond accuracy, including robustness, fairness, interpretability, and real-world utility.

AI Agents Interactive, goal-driven, and self-improving intelligent systems capable of operating independently or collaboratively in real and virtual environments.

Knowledge Automation: The process of encoding, updating, and utilizing domain-specific knowledge to automate reasoning, learning, and decision-making.

Autonomous Intelligence: The capacity of AI systems to sense, decide, learn, and act with minimal human intervention, enabling adaptive and scalable deployment across domains.

1.4.4 SE4SS: How New AI Solves Services Science Challenges

Services science faces key challenges due to its human-centered and cross-domain nature. Variability in individual emotions and perceptions leads to inconsistent service outcomes. Many processes remain manual, limiting automation and scalability.

Fig. 1.9 SE4SS: An experimental platform for new AI, scenarios engineering, and service systems

Data is often fragmented across platforms, hindering intelligent decision-making. A lack of standardized evaluation metrics makes performance hard to measure, and delivering personalized services at scale is difficult due to diverse user needs and limited contextual awareness.

SE4SS: A New AI Paradigm for Service Systems

Scenarios engineering delivers trustworthy AI by grounding the entire life cycle—design, training, calibration, certification, verification, and validation—within explicit, real-world contexts instead of isolated features. It uses the closed loop of I&I to set and quantify holistic goals such as safety, security, and sustainability, C&C to tune parameters against scenario data and issue blockchain-based performance certificates, and V&V to continuously confirm both "are we building the AI right?" and "are we building the right AI?". Because this scenario-driven loop is independent of any specific model architecture, it remains fully effective for new AI paradigms—whether future deep-learning variants, foundation models, or knowledge automation—by simply refreshing the scenario base and indexes without redesigning the framework itself.

In this context, new AI components offer a transformative solution to the enduring challenges in services science by introducing a more adaptive, intelligent, and scalable infrastructure. Therefore, we propose a general framework, called SE4SS (Scenarios Engineering for Services Science), for analyzing and guiding the development of service systems, as depicted in Fig. 1.9. The SE4SS framework facilitates the training, evaluation, and comparison of algorithms related to new AI, scenarios engineering, and service systems.

In SE4SS framework, scenarios engineering acts as a crucial bridge between technical advancements in AI and the goals of services science. This approach enables services science to move toward the realization of trustworthy and reliable service systems empowered by new AI. Specifically, service system plays a central role in addressing the key challenges of services science by leveraging the components of new AI. To address the human-centric variability inherent in service interactions,

complex scenario modeling allows for the simulation of diverse emotional, behavioral, and contextual factors, enabling AI systems to adaptively respond to human needs. **Foundation models**, trained on broad datasets, enhance the system's ability to understand and generalize across varied tasks, supporting consistent service quality despite individual differences. In response to low automation and limited scalability, **AI agents** and **autonomous intelligence** enable goal-driven, self-improving systems capable of automating complex workflows across dynamic environments. The problem of fragmented data is mitigated through **High-Quality datasets** and **knowledge automation**, which unify, curate, and operationalize data from disparate sources, enabling holistic analysis and real-time decision-making. **Advanced model evaluation metrics** offer robust frameworks to quantify service performance, even in subjective or experiential domains, overcoming the lack of standardized evaluation criteria. Finally, the challenge of personalizing services at scale is addressed by embedding **specific application scenarios** into AI systems, ensuring contextual awareness and task relevance, while **autonomous intelligence** supports continuous adaptation and responsiveness across individualized service demands. In summer, these components reframe service system as a data-rich, scenario-driven, and knowledge-automated discipline grounded in the realities of human-centric ecosystems.

Three Modes of New AI in Future Service Systems

In the future, new AI will bring unprecedented convenience to service systems, as illustrated in Fig 1.10. The morning phase will last over 20 hours and be dominated by Autonomous Modes (AM), during which robots and digital humans will automatically complete the tasks we need. This will be followed by the afternoon phase, lasting less than 3 hours, led by Parallel Modes (PM), where robots and digital humans perform tasks under remote human control and cloud-based guidance, continuing until evening. In the evening mode, we will enter the Expert/Emergency Mode (EM), which lasts no more than one hour. During this time, robots and digital humans will provide strong support, enabling humans to handle emergencies more easily.

1.5 Conclusion Remarks

Inspired by ACP, this chapter explores how scenarios engineering enables the development of trustworthy, adaptive, and human-centered new AI for service systems. While many contemporary AI approaches have achieved high levels of accuracy, they often do so at the cost of interpretability, transparency, and reliability—factors critical to user trust and system safety. These problems may easily lead to bad experiences, lower trust levels, and systematic or even catastrophic risks. We introduce the theoretical framework of Scenarios Engineering for building trustworthy AI techniques.

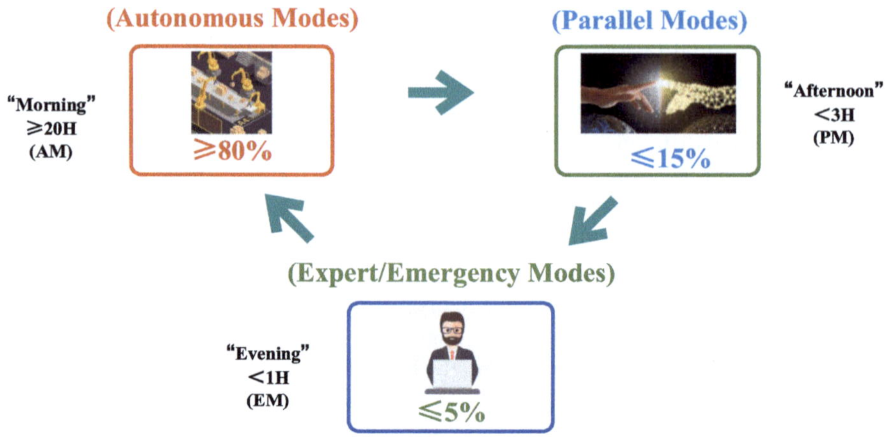

Fig. 1.10 A day in service systems transformed by new artificial intelligence: Three worlds, three modes

Moreover, the chapter defines new AI as a holistic, interdisciplinary ecosystem that extends beyond algorithmic optimization to include critical components such as complex scenario modeling, foundation models, high-quality datasets, autonomous agents, knowledge automation, and self-improving intelligence. These components collectively overcome traditional barriers in services science by enabling scalable personalization, adaptive service delivery, and data-driven decision support. Therefore, we propose a general framework, called SE4SS (Scenarios Engineering for Services Science), for analyzing and guiding the development of service systems. In SE4SS framework, scenarios engineering plays a central role in addressing key challenges in service systems by integrating components of the new AI.

Ultimately, scenarios engineering is positioned as a critical bridge connecting new AI technologies with real-world service applications. By reinforcing principles of transparency, interpretability, and continuous improvement, scenarios engineering advances the development of intelligent service systems that are not only technically proficient but also aligned with human ethics, social responsibility, and sustainability.

References

1. Daskin, M. S. (2011). *Service science*. John Wiley& Sons.
2. Qiu, R. G. (2014). *Service science: The foundations of service engineering and management*. John Wiley& Sons.
3. Wang, F.-Y. (2004). Artificial societies, computational experiments, and parallel systems: A discussion on computational theory of complex social-economic systems. *Complex Systems and Complexity Science, 1*(4), 25–35.
4. Wang, F.-Y. (2022). Parallel intelligence in metaverses: Welcome to Hanoi! *IEEE Intelligent Systems, 37*(1), 16–20.

5. Li, L., Lin, Y., Zheng, N., & Wang, F.-Y. (2017). Parallel learning: A perspective and a framework. *IEEE/CAA Journal of Automatica Sinica, 4*(3), 389–395.
6. Heaven, D. (2019). Why deep-learning AIs are so easy to fool. *Nature, 574*(7777), 163–166.
7. Li, X., Wang, Y., Yan, L., Wang, K., Deng, F., & Wang, F.-Y. (2019). ParallelEye-CS: A new dataset of synthetic images for testing the visual intelligence of intelligent vehicles. *IEEE Transactions on Vehicular Technology, 68*(10), 9619–9631.
8. Li, X., Wang, K., Tian, Y., Yan, L., Deng, F., & Wang, F.-Y. (2019). The ParallelEye dataset: A large collection of virtual images for traffic vision research. *IEEE Transactions on Intelligent Transportation Systems, 20*(6), 2072–2084.
9. Sun, C., Uwabeza, M., Li, Y., Chen, L., Li, L., Wang, F.-Y., Khajepour, A., & Cao, D. (2020). Proximity based automatic data annotation for autonomous driving. *IEEE/CAA Journal of Automatica Sinica, 7*(2), 395–404.
10. Wang, S., Housden, J., Bai, T., Liu, H., & Wang, F.-Y. (2021). Robotic intra-operative ultrasound: Virtual environments and parallel systems. *IEEE/CAA Journal of Automatica Sinica, 8*(5), 1095–1106.
11. Menzel, T., Bagschik, G., & Maurer, M. (2019). Scenarios for development, test and validation of automated vehicles. *IEEE intelligent vehicles symposium* (pp. 1821-1827).
12. Wang, F.-Y., Qin, R., Chen, Y., Tian, Y., Wang, X., & Hu, B. (2021). Federated ecology: Steps toward confederated intelligence. *IEEE Transactions on Computational Social Systems, 8*(2), 271–278.
13. Li, J., Wang, G., Wang, X., Li, J., Yuan, Y., & Wang, F. Y. (2022). Crypto management: A novel organizational management model based on blockchain. *Chinese Journal of Intelligent Science and Technology, 4*(2), 145–156.
14. Wang, S., Ding, W., Li, J., Yuan, Y., Ouyang, L., & Wang, F.-Y. (2019). Decentralized autonomous organizations: Concept, model, and applications. *IEEE Transactions on Computational Social Systems, 6*(5), 870–878.
15. Li, J., Yuan, Y., & Wang, F.-Y. (2021). Blockchain-based digital currency: The state of the art and future trends. *Acta Automatica Sinica, 47*(4), 715–729.
16. Wang, S., Ouyang, L., Yuan, Y., et al. (2019). Blockchain-enabled smart contracts: Architecture, applications, and future trends. *IEEE Transactions on Systems, Man, and Cybernetics: Systems, 49*(11), 2266–2277.
17. Wang, F.-Y., Wang, Y., Chen, Y., Tian, Y., Qi, H., Wang, X., Zhang, W., Zhang, J., & Yuan, Y. (2020). Federated ecology: From federated data to federated intelligence. *Chinese Journal of Intelligent Science and Technology, 2*(4), 305–311.
18. Liu, Y., Kang, Y., Xing, C., Chen, T., & Yang, Q. (2020). A secure federated transfer learning framework. *IEEE Intelligent Systems, 35*(4), 70–82.
19. Wang, F.-Y., Qin, R., Chen, Y., Tian, Y., Wang, X., & Hu, B. (2021). Federated ecology: Steps toward confederated intelligence. *IEEE Transactions on Computational Social Systems, 8*(2), 271–278.
20. Wang, F.-Y., Carley, K. M., Zeng, D., & Mao, W. (2007). Social computing: From social informatics to social intelligence. *IEEE Intelligent Systems, 22*(2), 79–83.
21. Wang, F.-Y., Wang, X., Li, J., Ye, P., & Li, Q. (2019). Social computing: From crowdsourcing to crowd intelligence by cyber movement organizations. *IEEE Transactions on Computational Social Systems, 6*(4), 619–626.
22. Wang, F.-Y., & Wang, Y. (2020). Parallel ecology for intelligent and smart Cyber–Physical–Social systems. *IEEE Transactions on Computational Social Systems, 7*(6), 1318–1323.
23. Li, X., Liu, Y., Wang, K., & Wang, F. Y. (2020). A recurrent attention and interaction model for pedestrian trajectory prediction. *IEEE/CAA Journal of Automatica Sinica, 7*(5), 1361–1370.
24. Zhang, X., Tao , J., Tan, K., Torngren, M., Sanchez, J., Ramli, M., & Felbinger, H. (2022). Finding critical scenarios for automated driving systems: A systematic mapping study. *IEEE Transactions on Software Engineering*, early access.
25. Ye, P., Chen, Y., Zhu, F., Lv, Y., Lu, W., & Wang, F.-Y. (2021). Bridging the micro and macro: Calibration of agent-based model using Mean-Field dynamics. *IEEE Transactions on Cybernetics*, early access.

26. Wang, F.-Y., Qin, R., Yuan Y., & Hu, B. (2021). Nonfungible tokens: Constructing value systems in parallel societies. *IEEE Transactions on Computational Social Systems, 8*(5), 1062–1067.
27. Klock, A., Gasparini, I., & Pimenta, M. (2016). 5W2H Framework: A guide to design, develop and evaluate the user-centered gamification. *Proceedings of the 15th Brazilian Symposium on Human Factors in Computing Systems* pp. 1–10.

Chapter 2
SE4SS for Intelligent Finance

2.1 Introduction

Due to the heavy reliance on human decision-making, traditional financial service system is facing challenges such as fluctuating service quality, high operational risk, and difficulty in expanding scale. Since the beginning of the twenty-first century, the rapid development of artificial intelligence technology represented by deep learning has brought great progress to the field of data science and analysis. The integration of services science and artificial intelligence is promoting the transformation of financial industry from process digitalization to scene intelligence. More and more advanced AI algorithms have penetrated into the financial service systems, including risk management, decision support, operation optimization, and process automation.

At the same time, the parallel intelligent computing paradigm proposed by Wang Feiyue at the beginning of the twenty-first century has won the favor of data scientists all over the world in the field of AI, and ACP (Artificial societies, Computing experiment, Parallel execution) has been introduced into modern AI research. The Scenarios Engineering (SE) of ACP provides a new paradigm for building a trustworthy intelligent financial system. Artificial societies (A): build a digital twin model of the financial market to simulate the multi-agent game environment. Computational experiment (C): verify the robustness of the model under extreme scenarios through virtual market test. Parallel execution (P): real-time coordination of decision optimization between real financial systems and artificial societies.

Based on the latest industry practice and research progress, this chapter systematically analyzes the technological breakthroughs and development trends in the field of Intelligent Finance from the dimensions of complex scenario generation, dataset construction, large model technology, knowledge automation, and fully autonomous services. Research shows that the development of financial AI has shifted from single point tool substitution to systematic business process reengineering. At the

X. Li and F.-Y. Wang, *Scenarios Engineering for Services Science*,
SpringerBriefs in Service Science, https://doi.org/10.1007/978-3-032-24152-8_2

same time, breakthroughs in reasoning accuracy and compliance reliability of large-scale professional financial models are pushing financial agents from technical concepts to large-scale implementation, reshaping the future of the financial service industry.

2.2 Paradigm Innovation in Financial Simulation

Financial risk management and market simulation mainly rely on the combination of classical mathematical models, statistical methods, and manual experience. Its core limitations are low computational efficiency, poor dynamic adaptability, and insufficient coverage of extreme scenarios. Traditional financial risk assessment relies on historical data and statistical assumptions, which is difficult to deal with the black swan event and cross market linkage effect. Generative AI-driven scene generation technology provides a breakthrough solution for risk management and decision-making training by synthesizing high fidelity market environment.

2.2.1 The Structures of Financial Data

Since the birth of computer technology and the popularity of machine learning methods, the way of realizing investment decision through mathematical models and algorithms has gradually replaced the traditional artificial investment relying on subjective judgment. The performance of these algorithms largely depends on the quality of training data. For quantitative investors, having more and higher quality datasets means being more forward-looking and leading in fierce market transactions. In the field of financial analysis, many forecasting algorithms, such as stock forecasting and quantitative trading, can use a variety of data forms.

Market data: including price information of financial products such as stocks, futures, foreign exchange, funds, etc., such as opening price, closing price, highest price, lowest price, trading volume, etc. This type of data is naturally time series data, usually presented in the form of multidimensional temporal signals.

Transaction data: mainly refers to order books and limit order books. Transaction data records specific transaction behavior, usually including timestamp, transaction price, transaction quantity, buying and selling direction, and other information. These data can also be converted into time series data for analyzing trading activity, price changes, etc.

Alternative data, such as social media emotions, and news reports, belong to unstructured data and usually require preprocessing before being converted into time series data. For example, sentiment indices can be extracted from social media data using natural language processing (NLP) techniques and recorded by timestamp.

Generally speaking, the above data can be collected from the company's financial reports, stock market websites, and financial news. However, in some important financial affairs, generating high-quality virtual financial history datasets is the key link of transaction strategy backtesting and AI training, which needs to take into account the statistical authenticity, market logic completeness, and extreme scenario coverage. Here are some examples.

2.2.2 MarS: Financial Market Simulation Engine

The MarS (Market Simulation System) [1] developed by Soft Asia Research Institute is based on a generative basic model, and its architecture is shown in the following figure. By learning the microstructure of the market through order level data, multiple sources of data such as stocks, bonds, and derivatives can be integrated to simulate the behavior patterns of market participants under different economic conditions. Its core innovation lies in adopting the Large Market Model (LMM) architecture to achieve controllable generation and dynamic evolution, providing a risk-free environment for strategy testing. The overview of the system is shown in Fig. 2.1.

MarS generates subsequent orders based on the company's recent orders and limit order book to simulate future market trajectories and achieve accurate market trend predictions. Specifically, MarS uses causal transformer to encode order sequences, ensuring that the context of each order is preserved across a wider range of marketing activities. Subsequently, an autoregressive transformer is applied to model the order batch sequence, capturing the structured patterns of the market within the aggregation time interval. Combining order sequence and order batch sequence modeling, balancing fine-grained control and broad market dynamics, to achieve improved market modeling and generation. Meanwhile, MarS provides a fine-grained signal generation interface that maps fuzzy descriptions from natural language or general configurations to fine-grained control signal sequences, ensuring that simulations follow real-world market patterns. After training, MarS generates a real order flow and provides the necessary information for subsequent order generation.

2.2.3 Generative Methods for Market Simulation

In order to ensure security, investment banks and trading companies now prefer to use a simulated environment to evaluate trading strategies before trading in the real market. This technique of reproducing trading strategies under various existing historical scenarios referred to as backtesting. However, the back testing technology only considers the market interaction behavior of experimental agents, which is difficult to capture the market response to the behavior of multi-agent, and lacks the interaction between experimental agents and other agents, making the effect of simulated trading strategy unreliable. Therefore, the establishment of a market simulation system with multi-agent interaction has become a research hotspot in the financial field.

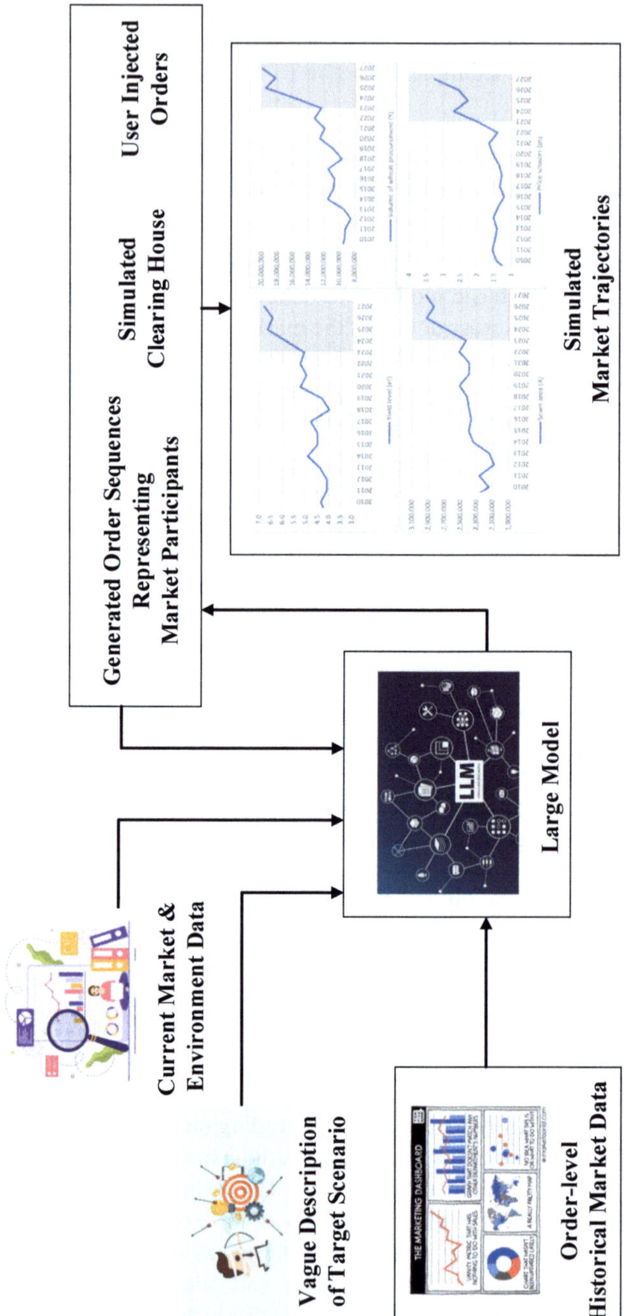

Fig. 2.1 Overview of MarS

Andrea coletta et al. proposed a data-driven and responsive market simulation framework for the evaluation and training of trading strategies. The core technology of the framework is to generate synthetic market orders based on conditional generative adversarial networks (CGAN), which aims to create a simulation environment that can reflect the real behavior of the market and respond to experimental strategies. Fig. 2.2 shows its flow. The framework consists of four parts: (1) Exchange: it is responsible for processing all incoming orders, including limit orders and market orders. (2) Experimental agent: an agent used to test specific trading strategies. (3) CGAN-based world agent: the agent based on CGAN is responsible for simulating the behavior of the whole market and generating new orders according to the current market state. (4) Historical data: used to train the CGAN agent and initialize the simulation environment.

Different from other multi-agent simulation technologies, the framework innovatively uses generative method to represent the market behavior of multi-agent abstractly. At the same time, the conditional generation countermeasure network is used to ensure the reliability and controllability of the generated simulation data, which corresponds to the requirements of Scenarios Engineering for model credibility.

2.3 Foundation Models Creates Financial Paradigms

The response of modern financial market to all kinds of information is extremely rapid. Whether it is the release of economic data at the macro level or enterprise news at the micro level, it may cause market fluctuations. The efficient market hypothesis

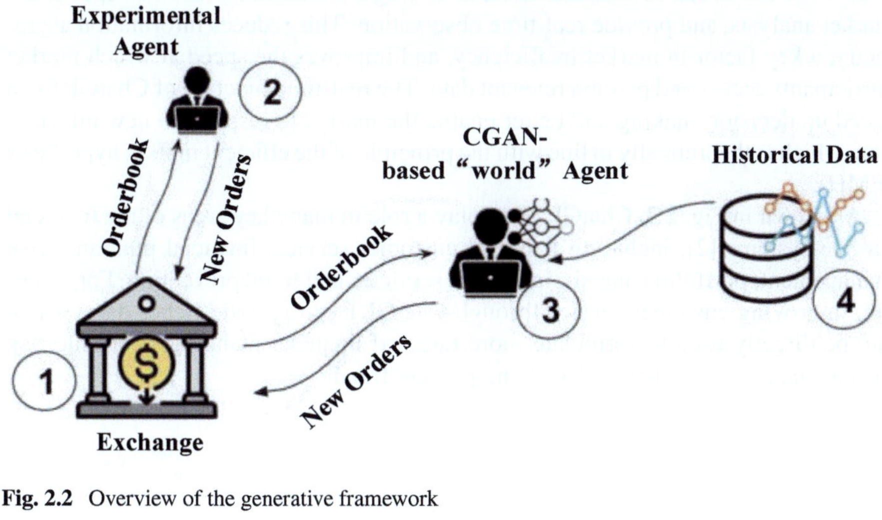

Fig. 2.2 Overview of the generative framework

(EMH) holds that the financial market is "information efficient," which means that asset prices can fully reflect all available information at any particular time. This means that in an efficient market, it is impossible to continuously obtain higher returns than the overall market without taking additional risks. Prices should be adjusted immediately based on new information, which will be reflected in asset prices.

This has triggered thinking on the development of the financial industry in the AI era. How to build a perfect model that can handle and integrate highly structured complex market information and bring higher reliability and efficiency to financial decision-making.

2.3.1 ChatGPT Applications in Intelligent Finance

Large language model (LLM) is an artificial intelligence system based on deep learning. It is trained by massive text data and can understand, generate, and reason human language. Its essence is to build a probability machine that can simulate the human language model, and grasp the language rules by predicting the relationship between words in the context. Large language models are usually pre-trained through large-scale datasets to learn the general knowledge base, and can fine-tune tasks in specific fields through migration learning. Among them, ChatGPT proposed by openai adopts transformer decoder only architecture, which can capture complex language patterns and dependencies and attracts much attention.

ChatGPT is an advanced artificial intelligence model with complex natural language processing capabilities. It has the potential to improve market efficiency by improving information dissemination and accessibility. ChatGPT can analyze a large number of unstructured financial data, including news articles, income reports and market analysis, and provide real-time observation. This reduces information asymmetry, a key factor in market inefficiency, and improves the speed at which market participants access and process relevant data. The real-time function of ChatGPT can speed up decision-making and better enable the market to respond to new information, which is theoretically in line with the principle of the efficient market hypothesis (EMH).

As shown in Fig. 2.3, ChatGPT can play a role in many key areas of the financial service systems [2], including financial customer service, financial planning, risk management, portfolio analysis, insurance services, and fraud prevention. For example, improving customer service through ChatGPT can provide richer data, which can be directly used to formulate more targeted financial planning; real-time risk management can also better detect and prevent fraud.

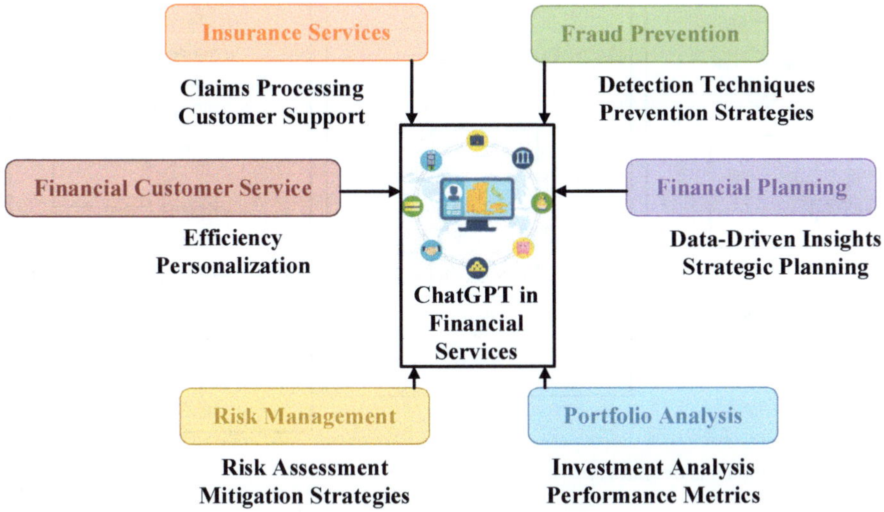

Fig. 2.3 ChatGPT's core application business in the financial field

2.3.2 Personalized Customers of Financial Service Systems

The contemporary financial service system is undergoing a profound change driven by large language models. The financial service model is accelerating the transformation from the traditional paradigm of standardization and batch to the direction of highly personalized and intelligent. The core of this transformation is that financial agents based on the large language model can realize the in-depth understanding and prediction of customer needs through independent perception, decision-making, and execution ability, so as to provide accurate services for "thousands of people and thousands of faces." Compared with traditional AI tools, such agents are no longer simple procedures for passive response to instructions, but active service agents with business understanding, situational analysis, and real-time interaction capabilities, fundamentally changing the way financial institutions connect with customers.

Tao Xu et al. proposed the autonomous fundamental analysis system (Auto-FAS) [3]. The inspiration of the system originates from the process of human analytical knowledge and accumulation. Through the iterative generation of research memos, LLM agents have made significant improvements in the performance of long-term and short-term factor analysis and unique perspectives compared with the general model.

The system is composed of three units: analysis unit, novice agent, and expert agent. The analysis unit is responsible for intelligent selection of required materials from the document directory and specific analysis tasks. Novice agent is the initialization of expert agent. It can be self-developed into expert agent through multi-angle analysis through the cycle of "think assign task complete." Expert agent is upgraded from novice agent after completing a certain number of analyses. It holds "research

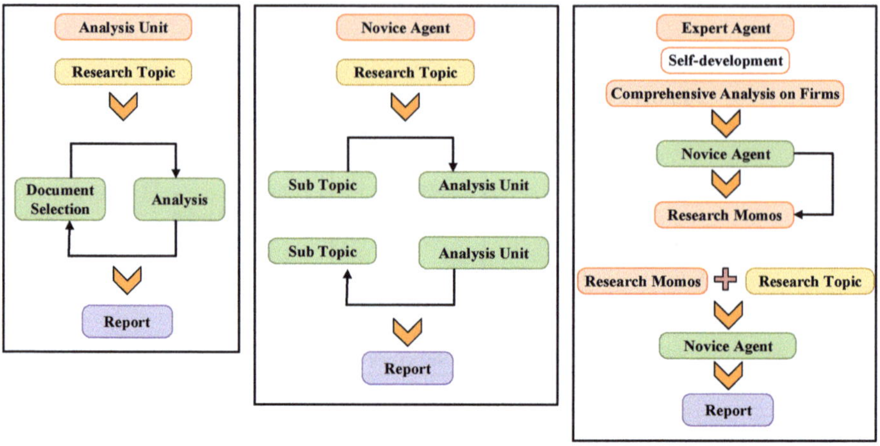

Fig. 2.4 Components of AutoFAS

memos" and refers to these memos in new tasks to generate more in-depth and unique reports. Fig. 2.4 shows its overview.

The design of the model takes GPT-4o from OpenAI as the underlying basic model, and the data comes from the quarterly financial reports, financial reports, and conference call records of several companies in Japan's five major industries (automobile, steel, business, telecommunications, and insurance) from January 2022 to June 2024. In terms of model evaluation, the experiment uses 8 evaluation criteria (such as quantitative analysis, long-term factors, unique perspectives, insights, etc.), the model generates 100 research topics, and novice and expert agent generate reports, respectively. Finally, three human analysts are combined for comparative evaluation.

2.4 Intelligent Autonomy for Finance

With the in-depth development of artificial intelligence technology, AI agent technology has been proposed in recent years and shines brightly. At the same time, the concept of AI agent is profoundly reshaping the underlying architecture and operation paradigm of modern financial service systems. The limitations of traditional service systems relying on static rule engines and manual decision-making have become increasingly prominent, especially when dealing with high-dimensional dynamic financial markets.

The penetration of artificial intelligence agent (AI agent) in the field of financial service system has gone beyond the level of technology application and is becoming a transformative paradigm for reconstructing the theory and practice of finance. This process is essentially the product of the deep integration of multi-agent system (MAS) theory and complex financial ecosystem. Its academic value is reflected not only in the improvement of operational efficiency, but also in the theoretical expansion and methodological innovation of the three pillars of traditional finance: risk management, asset pricing, and market microstructure. The current academic research focuses on how the autonomous decision-making mechanism of agent solves the sequential decision-making problem in the financial field through reinforcement learning (RL), multi-task learning (MTL), and other frameworks, and at the same time leads to deep speculation on algorithmic governance and financial stability. How to build an efficient and accurate AI agent with professional economic knowledge is an important proposition in the field of AI for intelligent finance.

2.4.1 AI Agents for Economic Strategy Formulation

Stephan Zheng et al. [4] proposed an AI economist framework based on two-level deep reinforcement learning. The framework combines machine learning with AI-driven economic simulation to overcome the limitations of existing methods and demonstrates the effectiveness and feasibility of using two-level reinforcement learning as a new paradigm for economic policy design. Its overview is shown in the figure.

In the framework, AI social planners (governments) at the upper level optimize social welfare objectives (such as the balance between equality and productivity) by adjusting tax policies (such as marginal tax rates), while heterogeneous RL agents (residents) at the lower level learn utility maximization behaviors such as labor, trading, and construction under the constraints of tax policies. The coadaptation of the two is the key innovation: policy changes affect agent behavior, and behavior feedback drives policy optimization to form a dynamic closed loop. In order to overcome the instability of training, the researchers developed structured course learning (gradually transition from tax-free environment to full tax environment) and entropy-based regularization (forcing the strategy to maintain randomness and avoid premature convergence), which successfully solved the problem of "Lucas' criticism" faced by the traditional methods because of ignoring behavioral feedback.

The framework combines machine learning and AI-driven economic simulation to test and optimize policies by simulating complex economic environments, as shown in Fig. 2.5. Different from the traditional agent-based model, the agent behavior in AI economist is not pre-defined, but dynamically emerged through deep reinforcement learning, which can more truly reflect the economic phenomenon in the real world. This method shows the feasibility of using deep RL for economic policy design, and provides a new computing method and AI-driven framework for economic policymaking, which can deal with the complex economic environment that traditional methods are difficult to solve.

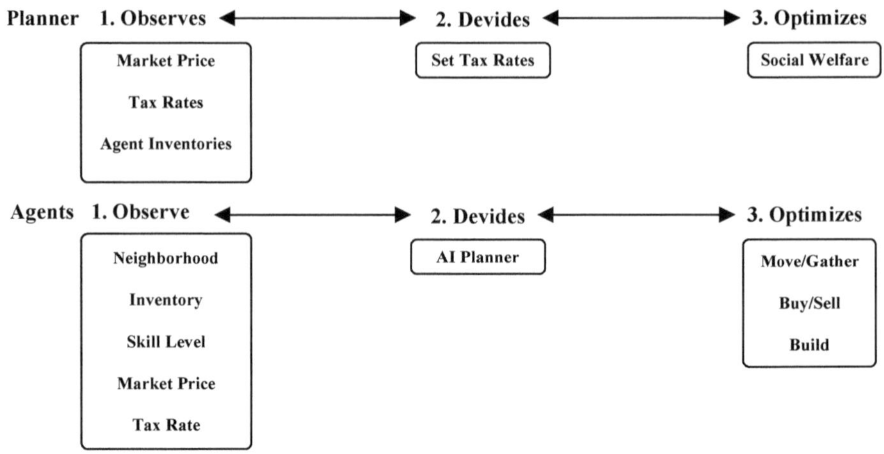

Fig. 2.5 Framework of two-layer deep reinforcement learning for AI Economist

2.4.2 Multi-Agents Market Systems

Based on the discrete event simulation (DES) framework, David Byrd et al. [5] combined the characteristics of multi-agent system (ABMS) and proposed Agent-Based Interactive Discrete Event Simulation (ABIDES). It aims to simulate the complex dynamics in the real market. It provides a virtual environment to achieve platform independence and can quickly reconfigure different agent groups, market conditions, exchange rules, and agent superparameters.

ABIDES' architecture includes a customizable configuration system, a simulation kernel, and a rich hierarchy of agent functions. The kernel is responsible for processing all agent messages, supporting the simulation of geography, computing time and network latency, and ensuring the authenticity of the simulation. Agent class is the basic class that all participants must inherit. It implements the methods required to participate in the simulation life cycle. This design makes it not only serve Financial Research (such as simulating market shock, CO location value evaluation, algorithm strategy testing), but also can be extended to other fields. For example, in the simulated secure multi-party federated learning protocol, the distributed client is modeled as an agent, network communication, and differential privacy noise injection are simulated in the kernel, and the impact of different size (n) and privacy budget (ε) on the accuracy of the model is evaluated.

2.5 Conclusion Remarks

This chapter explores the role of scenarios engineering technology in the financial field and deeply analyzes the changes and values it brings at multiple key levels. In today's financial investment field, including stock investment and quantitative trading, with the popularization of machine learning technology, there is a gradual shift toward automated decision-making. However, such algorithms still face limitations. Firstly, they require a very large amount of data as performance support. Secondly, current artificial intelligence technology still does not have good interpretability, and its security and credibility are difficult to guarantee. Therefore, the concept of financial scenarios engineering emerged naturally, and its core value lies in risk simulation and decision support.

This chapter explores the positive role of new AI in promoting intelligent finance from the perspectives of complex scenario generation, large model-assisted decision-making, and AI agents. It also lists some advanced artificial intelligence models and their outstanding performances for financial decision-making and risk management.

Overall, intelligent finance is not only a powerful efficiency tool and innovation catalyst, but also a key driving force for promoting the intelligence, personalization, and refined risk management of financial service systems. Its continuous evolution and responsible application will continue to drive the intelligent upgrading of the financial industry in terms of efficiency, insight, and resilience.

References

1. Li, J., Liu, Y., Liu, W., Fang, S., Wang, L., Xu , C., & Bian, J. (2025). MarS: A financial market simulation engine powered by generative foundation model. In *The thirteenth international conference on learning representations*. Early access.
2. Chen, Z. (2025). Revolutionizing finance with conversational AI: A focus on chatGPT implementation and challenges. *Humanities and Social Sciences Communications, 12*(388), 1–11.
3. Xu, T., Piao, Z., Mukai, T., Murayama, Y., & Izumi, K. (2025). *Can large language models autonomously generate unique and profound insights in fundamental analysis?* Early access: Intelligent Systems with Applications.
4. Zheng, S., Trott, A., Srinivasa, S., Parkes, D. C., & Socher, R. (2022). The AI economist: Taxation policy design via two-level deep multiagent reinforcement learning. *Science Advances, 8*(18), 1–17.
5. Byrd, D., Hybinette, M., & Balch, T. H. (2020). ABIDES: Towards high-fidelity multi-agent market simulation. In *Proceedings of the 2020 ACM SIGSIM conference on principles of advanced discrete simulation* (pp. 11–22).

Chapter 3
SE4SS for Intelligent Customer Services

3.1 Introduction

In today's world where business competition is increasingly focused on user experience, customer service has jumped from a simple cost center to a key battlefield for enterprise value creation. From the traditional model of relying on human communication to the AI-driven intelligent customer service system, this evolution has profoundly reshaped the way businesses interact with users.

Traditional customer service takes manual agents as the core and provides point-to-point service through telephone, email, and other channels, and its greatest value lies in the temperature and flexibility of the "people." Experienced customer service representatives can keenly capture user emotions and show irreplaceable advantages in complex and non-standardized scenarios. The natural emotional connection between people has a natural power in building trust. However, the limitations of the traditional model are becoming increasingly prominent: high labor costs, limited service time, inefficiency in the face of massive repetitive consultations, long waiting times for users, and poor experience. Information silos are common, and users often need to state their problems repeatedly. These pain points have become the driving force behind the change in the customer service model.

The rise of intelligent customer services is essentially the deep integration and application of artificial intelligence, big data, and other technologies in the field of customer service. It mainly exists in the form of virtual assistants and intelligent voice response systems, providing round-the-clock services through multiple touchpoints. Its groundbreaking innovation lies in the exponential improvement of efficiency. Intelligent customer services can process massive standardized consultations in parallel in an instant, reducing the average waiting time of users to almost zero. The accessibility of the service has been greatly expanded, allowing users to get instant responses anytime, anywhere. The system can also continuously analyze interaction data to reveal user behavior patterns and provide valuable insights for companies to optimize their products.

X. Li and F.-Y. Wang, *Scenarios Engineering for Services Science*,
SpringerBriefs in Service Science, https://doi.org/10.1007/978-3-032-24152-8_3

Despite significant advantages, the development of intelligent customer services still faces challenges. Current technology has limited ability to deal with highly complex and ambiguous problems, and dialogue sometimes seems mechanical or even "answers to questions"; over-reliance on automation can weaken the "human touch" of services, especially in sensitive scenarios where emotional support is needed; ethical issues such as algorithmic bias and data privacy breaches cannot be ignored [1].

Scenarios Engineering for Services Science (SE4SS) is a services science methodology that integrates scenarios engineering and new artificial intelligence technology. In this chapter, we aim to apply SE4SS to build a highly reliable, flexible, and user-centered artificial intelligent customer service systems. This methodology systematically constructs multi-level service scenarios (including user roles, behavior patterns, environmental variables, and contacts), combining deep learning, real-time decision support, and continuous learning mechanisms to ensure that intelligent customer services maintain service stability and user experience in complex and dynamic environments. The core of SE4SS is to use scenarios as the "strategic link" of service design: on the one hand, through application-oriented scene visualization, such as user journey diagrams and future scenario simulations, the contacts and fault boundaries in the service process are accurately defined; on the other hand, with the help of AI-driven dynamic scene evolution, such as sequence-to-sequence dialogue generation, multimodal intent recognition, and fault prediction, the active adaptation and self-healing ability of services are realized.

This chapter mainly introduces SE4SS based on new artificial intelligence for intelligent customer services from the following aspects: complex scenarios; high-quality datasets; foundation model; advanced model evaluation metrics.

3.2 Intelligent Customer Services Scenarios

The evolution of intelligent customer service requires navigating between two interconnected yet distinct realms: real-world scenarios rooted in human experience interactions, and synthetic scenarios designed to simulate complexity beyond the limits of observation. In the intelligent customer service systems, real scenario data and generated scenario data constitute a key dual-track driving mechanism, and the two jointly promote the continuous evolution of the intelligent level of the system through functional complementarity and collaborative iteration. This duality of complex scenarios connecting the real world with the generative world forms the core framework of SE4SS, which verifies its performance in real-world chaotic environments and synthetic scenarios by accurately testing AI-driven reliability.

3.2.1 Real Customer Services Scenarios

A real-world scenario is a dynamic environment composed of noise observation data, partially observable states, and open-world assumptions, characterized by unpredictability, cultural differences, and high-risk decisions. The core value of real-world scenario data comes from the historical interaction records between users and the customer service system, and its core value is to objectively reflect the distribution of users' actual needs, natural language expression patterns, contextual association characteristics, and pain points in the service process. This dataset directly supports the training and verification of the core components of intelligent customer services: it is used as the basic sample for supervised learning to train the intent recognition, semantic parsing, and dialogue state tracking models. The accurate construction and dynamic optimization of the knowledge base are driven by the analysis of high-frequency queries and the distribution of long-tail problems, evaluate the validity of the answers in the knowledge base based on the user's subsequent interaction behavior, identify process design defects with the help of dialogue path analysis, and optimize the multi-round dialogue strategy and manual decision-making mechanism. Real data also provides an initial sample reference for the cold start of new services, but is limited by insufficient long-tail coverage, lagging business changes, and data privacy constraints.

The authenticity of real-world scenarios is reflected in the complexity that cannot be reduced to a single element—both chaotic and contradictory, difficult to standardize. Customer service AI needs to identify key intents from vague and emotionally charged conversations. In terms of emotion recognition, a hybrid model combining Convolutional Neural Network (CNN) and Long Short-Term Memory Network (LSTM) is usually used to evaluate its performance through a variety of physiological parameters [2], and the model framework for recognizing emotions based on CNN and LSTM is shown in Fig. 3.1. Conversational AI in the medical field must interpret the patient's symptoms spoken while following the doctor's instructions. For example, a patient may say "My headache is like a drum," which is not found in clinical textbooks, but contains rich semantic urgency. Traditional rule-based systems fail in such contexts because they cannot understand metaphorical contexts or differentiate between emergencies (e.g., migraine and stroke).

Real-world scenarios require the system to be dynamically adaptable: when the system detects frustration in the customer's voice, it must escalate the issue to human customer service without adding tension. This requires the system to have multimodal perception capabilities—integrating information such as speech timbre, vocabulary selection, and even silence duration.

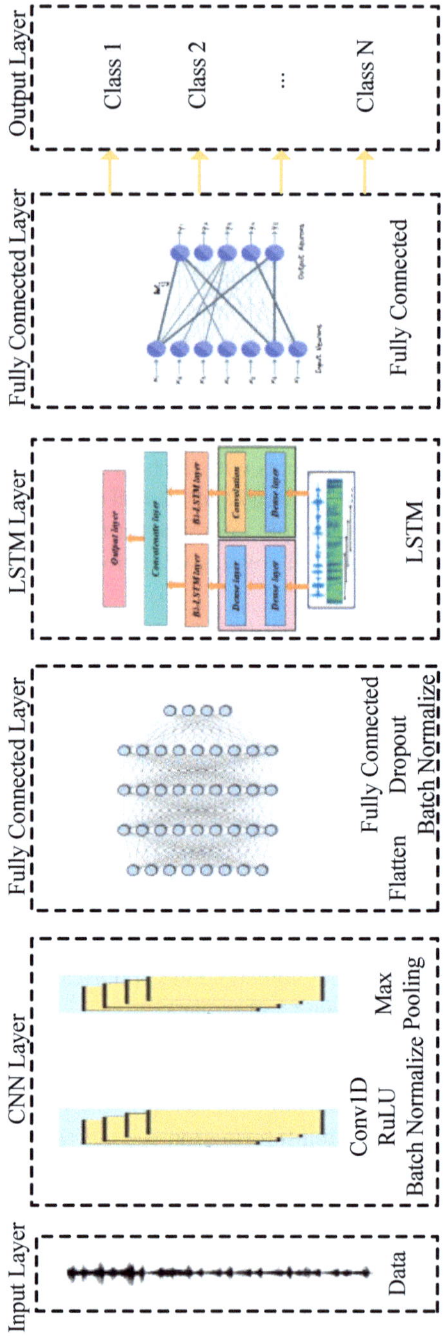

Fig. 3.1 Model framework based on CNN and LSTM to identify emotions

3.2.2 Synthetic Customer Services Scenarios

Synthetic scenario data is synthesized by large language model technology, and simulated dialogues are constructed based on existing knowledge bases, product documents, and a small number of real samples. Synthetic scenarios are designed to test the system's ability to respond to rare but catastrophic failures. These scenarios do not simply mimic the real world, but are algorithmically generated high-dimensional spaces that simulate service scenarios in human interactions that are too rare, dangerous, or complex to fully observe in the natural environment. The core of the generation scenario is to break through the data scarcity and ethical boundaries of the physical world, and build an "anti-stress immune system" for the artificial intelligence system: systematically generate low-frequency long-tail problem samples to make up for the distribution bias of training data; rehearse new business scenario dialogues to solve the data vacuum problem in the cold start stage; customize the creation of extreme scenarios and diverse language variants to enhance model robustness; avoid sensitive information and meet data compliance requirements with controllable generation mechanisms.

The rise of Generative Artificial Intelligence (GenAI) has injected disruptive momentum into the synthetic scenario paradigm. Large language models represented by Generative Pre-trained Transformer 4 (GPT-4), with their powerful pattern recognition, knowledge association, and creative text generation capabilities, can foresee complex failure chains that human engineers may overlook with unprecedented scale and diversity. They can simulate unusual combinations of physical laws, accidental couplings of socio-technical systems, and even malicious attack vectors that have never appeared in reality. This capability makes GenAI a "super engine" for generating high-fidelity, high-complexity synthesis scenes. The GPT-4 architecture enables the construction of synthetic user personas with multi-layered psychological characteristics that interact in the digital twin of the service ecosystem, revealing hidden failure modes: exposing the vulnerability of decentralized AI to data poisoning attacks [3].

When using AI to generate synthetic scenarios, such as virtual environments, training data, and simulated conversations, the core risk is falling into the "simulation trap." This phenomenon refers to the overfitting of the distribution law of the training data by the generative model, resulting in the output scene, although self-consistent and logical, seriously deviates from the complexity and constraints of the real world, and becomes a refined "simulation illusion." These scenarios lack deep embedding of real physical laws, social norms, ethical boundaries, or legal frameworks, such as generating object trajectories that perfectly conform to physical laws but cannot actually exist, or fictional logical interactions that violate data privacy regulations such as the General Data Protection Regulation (GDPR). This trap of "overfitting generative data" departs synthetic scenarios from their reliability as a basis for effective real-world simulations or training, potentially leading to misleading and even dangerous results for downstream applications.

To solve this problem, the hybrid architecture called ChemCrow offers an effective solution, which integrates the creativity of Large Language Models (LLMs) with the rule-binding capabilities of symbolic AI [4]. Under this framework, LLMs are responsible for coming up with a wide variety of initial scene ideas, describing details, and driving narrative possibilities. At the same time, symbolic AI performs real-time or iterative verification, correction, and constraint enhancement of scenario proposals generated by LLMs based on clear and interpretable rules, knowledge bases (such as physical laws, chemical principles), and hard constraints (such as GDPR and other regulatory provisions, security standards, and industry norms).

3.3 High-Quality Customer Services Datasets

3.3.1 Definition and Structure of Datasets

Driven by artificial intelligence and natural language processing technology, intelligent customer service systems have become an important hub for enterprises to interact with users. One of the core elements supporting its development is the intelligent customer services dataset. Datasets are not only fuel for training models but also yardsticks for evaluating system performance. This type of dataset is centered on multiple rounds of human-machine dialogue, systematically recording user consultations, customer service responses, and related interaction metadata, aiming to enable machines to understand user intentions, generate appropriate responses, and complete specific tasks. Liu et al. [5] emphasize that the construction of high-quality conversational datasets needs to reflect the complexity of real interactions, including noise, ambiguity, and multiple rounds of contextual dependencies.

The structure of the intelligent customer services dataset presents multi-dimensional characteristics, and its basic component unit is the dialogue round, each of which contains user input and customer service response. User input often contains colloquial expressions, misspellings, and ambiguous descriptions, which require real linguistic complexity in the dataset. The system response needs to cover a variety of strategies, from direct answers to rhetorical clarifications. Datasets also often contain rich metadata tags: user intent, conversation state, emotional tendencies, entity information, and more. As a typical representative, the DialoGLUE dataset integrates multiple conversational tasks, such as intent recognition, slot filling, and response generation, and its multitasking characteristics highlight the multi-label and structured trend of modern intelligent customer services datasets [6].

3.3.2 *Dataset Sources and Challenges*

There are two main sources of data for intelligent customer services: public resources and private logs of enterprises. Public resources such as the Dialog State Tracking Challenge (DSTC) series provide standardized benchmarks but may lack business relevance. Although enterprise private logs are closer to the real world, they face strict data privacy compliance risks, and need to use advanced desensitization technologies such as differential privacy and entity replacement to meet regulatory requirements such as GDPR.

The preprocessing process involves complex text cleaning, anonymization, and dialogue segmentation, and data bias problems such as the high proportion of specific user groups or problem types often need to be corrected by resampling [7] or synthetic data technology, as shown in Fig. 3.2. Manual annotation is the core guarantee of data quality, and well-trained annotators add labels such as intent and slots according to specifications, and data quality relies on strict multi-person cross-verification and statistical index review.

In addition, difficulties in domain migration, such as the difficulty of adapting models trained on e-commerce data to financial customer services and the lack of support for multilingual scenarios, further increase the complexity of building general datasets.

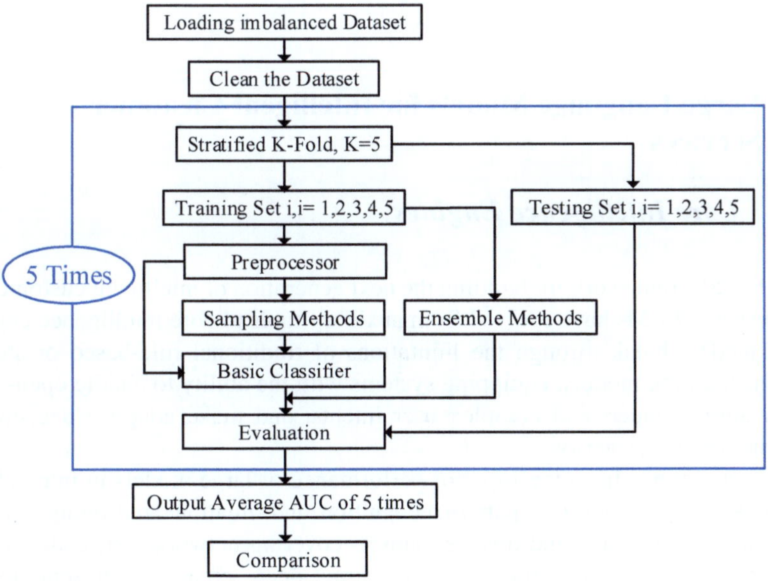

Fig. 3.2 Resampling flow chart for unbalanced data

3.3.3 Dataset Applications in Service Life Cycle

The application of intelligent customer services datasets runs through the entire life cycle of the system, and its effectiveness is limited by data characteristics and environmental changes.

In the model training and optimization stages, intent recognition, conversation state tracking, and end-to-end generation models all rely on large-scale, high-quality data. The emotionally enhanced customer service system trains a multimodal analysis module through conversation datasets to identify users' emotional states (such as anger and anxiety) in real time, and dynamically adjusts response strategies accordingly, such as triggering a priority response mechanism when high anger values are detected.

In system evaluation and iteration, datasets provide objective metrics such as intent recognition accuracy and response generation scores, but static test sets are difficult to reflect the dynamic evolution of the online environment, and new user questions, emerging terms, or emergencies may lead to model failure. Although the sentiment analysis module of dataset training can assist in strategy optimization, it is susceptible to cultural differences and expression diversity. In addition, when using historical dialogues for business insight mining, data sparsity may affect the reliability of conclusions. After the system was launched, continuous data collection and incremental learning became a rigid need to cope with changes in language habits.

3.4 Large Language Models for Intelligent Customer Services

3.4.1 Core Intelligence Engines

In the SE4SS framework for building the next generation of intelligent customer service systems, LLMs have evolved from auxiliary tools to core intelligence engines. These models break through the limitations of traditional rule-based or shallow machine learning models, equipping systems with the ability to handle open-ended conversations, understand complex user intents, and make adaptive decisions in dynamic service scenarios.

This shift stems from the superior performance of large models in internalizing massive knowledge, complex pattern recognition, and situational reasoning. Through pre-training on massive and diverse datasets covering network text, code, conversation records, etc., large models can construct deep semantic representations of human language, common sense, and domain knowledge. This allows them to not only identify implicit context, emotional nuances, and even cultural context differences in user queries, far beyond simple keyword matching. When users express emotional feedback on product issues, large models can not only identify the core problem of product failures, but also capture the intensity of user negative emotions and correlate potential solution paths (e.g., returns, exchanges, technical support). This deep understanding forms the cornerstone of delivering precise, personalized service. The large model elevates customer service from mechanical response to deep understanding, laying a cognitive foundation for fully autonomous service. DeepMind's AlphaFold system has used large models to resolve the complex semantics of protein sequences, overcoming long-standing problems in the field of biology [7]; this deep semantic understanding is also applicable to user intent parsing in customer service scenarios.

Knowledge bases alone are no longer sufficient to handle the complexity of real-world customer scenarios. Contextual learning and generalization capabilities are the key mechanisms for large speech models to demonstrate strong adaptability in SE4SS. Unlike traditional models that require extensive retraining or fine-tuning for each new task, large language models can quickly adapt to new tasks with a small number of samples (small-shot learning) or even just from task descriptions (zero-shot learning). This feature is particularly important for customer service scenarios—new situations such as new product launches, changes in promotion rules, and public relations events in emergencies are emerging. Systems based on large language models only need to integrate the latest policy documents and a small number of dialogue samples to understand the special process under "force majeure" and generate personalized solutions that comply with the new regulations, without the need for weeks of model retraining. This dynamic adaptability significantly improves the system's stress resistance and response speed. The Pathways Language Model (PaLM) demonstrates this excellent few-shot and zero-shot generalization capabilities, solving novel and complex reasoning problems based on task descriptions alone [8], which is the core requirement for dealing with dynamic customer service scenarios.

3.4.2 *Multimodal Fusion and Complex Reasoning*

Multimodal large language models use a unified learning framework to semantically align text, images, speech, and even structured data (such as user historical orders and knowledge base entries) to the same high-dimensional space. When users send graphic information at the same time, the model can synchronously parse image content and text information, combine user data, and automatically start product processing processes at different levels according to image characteristics and emotional states. In the SE4SS system, this means that the customer service system can truly "understand" user demands, achieving more accurate scene judgment and service triggering.

The core breakthrough lies in advanced reasoning technology driven by large language models. Customer service issues often involve multi-step logic, which requires the system to be able to handle multiple issues simultaneously. Chain of thought reasoning technology guides the model to explicitly decompose problems, which significantly improves its ability to solve complex problems. The tool call function makes the model a "command center," which can determine when to query the database, call a computing tool or assign a task, and integrate the results to generate a coherent response. The neural symbol combination method combines the pattern recognition of neural networks with the accurate reasoning of symbolic logic and greatly improves the interpretation ability and complex problem handling ability [9].

3.4.3 Key Application Scenarios

Seamlessly and reliably integrating the capabilities of foundation models into real-world customer service systems presents a serious systems engineering challenge, far from simple model deployment, but rather a complex engineering system that involves architectural design, trusted assurance, and continuous evolution.

Modular architecture design is key to balancing performance and efficiency. A single large model struggles to meet the customer service needs of high-frequency interactions in terms of response latency and computational cost. An effective strategy is to adopt a "large model and small model collaboration" architecture: lightweight models handle high-frequency simple queries. Automatically switch to a more powerful large or specialized model when complex intent is detected; another trend is the Mixture of Experts (MoE) architecture, where different expert subnetworks within the model are dynamically activated according to task requirements, as shown in Fig. 3.3.

Trustworthiness and accountability form the lifeline of foundation model deployment in customer service, and its key challenges run through multiple key dimensions.

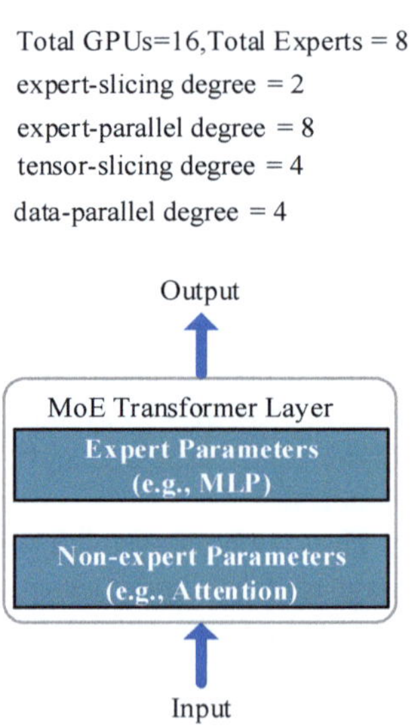

Fig. 3.3 DeepSpeed-MoE design that embraces the complexity of multi-dimensional parallelism for different partitions (expert and non-expert) of the model

The hidden social bias in the training data may trigger discriminatory responses, which requires the formation of a combined defense mechanism through counterfactual data enhancement, fairness constraint training, and output filtering. At the same time, the basic model may generate fictional information with a highly confident attitude, and it is necessary to build a Retrieval-Augmented Generation (RAG) architecture to strictly anchor the output to authoritative knowledge sources, supplemented by a fact consistency verification module for double guarantee. When it comes to contentious decisions, the system must provide an understandable basis for reasoning, which can be reinforced by visualizing attention mechanisms and generating decision-based text. Finally, to prevent the generation of harmful, misleading, and unethical content, it is necessary to establish a multi-level protection system: the front-end deploys input sensitive word filtering, the back-end implements output content security scanning, and uses value-based intensive learning and training to build ethical guardrails, so as to form a closed loop of responsibility throughout the whole process.

Continuous learning and service evolution are core requirements of SE4SS. Customer service scenario knowledge such as products, policies, and user preferences are constantly changing, and the foundation model needs to support safe and efficient incremental learning or efficient parameter fine-tuning to avoid catastrophic forgetting. Combined with Reinforcement Learning (RL), the system can automatically optimize dialogue strategies based on user satisfaction feedback such as ratings and complaint rates. At the same time, it is essential to establish a strict model evaluation and monitoring system, which needs to go beyond traditional metrics and pay attention to business value and potential risk metrics.

3.5 Customer Services Model Evaluation Metrics

3.5.1 Basic Evaluation Metrics

The core value of the intelligent customer service system lies in its ability to accurately and efficiently solve user problems. Accuracy and task success form the primary dimension of evaluating this core competence. Accuracy is typically applied to tasks with clear standard answers, such as intent recognition, entity extraction, and Frequently Asked Questions (FAQ) matching, to measure how well the model output aligns with preset criteria. However, in dynamic, multi-round conversational interactions, the accuracy of literal matching is often not enough to reflect real service performance. The task completion rate goes a step further, focusing on whether the user's core demands are substantively addressed at the end of the conversation, such as whether the order modification was successfully completed and whether specific policy questions were clearly answered. Therefore, defining fine-grained task success criteria based on specific business scenarios is a key prerequisite for accurately evaluating the practicability of the model.

Efficiency metrics are a direct bridge between system performance and user experience and operating costs. Response time, that is, the time delay from the user's initiation of the query to the system's first valid reply, is a key factor affecting user satisfaction, especially in mobile instant interaction scenarios, where millisecond differences can significantly change the user's perception of service responsiveness [11]. For example, the service contact satisfaction (SAT) given by the elderly and young people for different response times is shown in Fig 3.4. Conversation rounds quantify the average number of interactions the system needs to complete a typical task. A lower round usually means that the model has stronger intent understanding, information integration ability, and proactive service awareness, which can more efficiently guide the conversation to the target completion point, thereby reducing the operational burden on users and improving the overall service fluency.

Robustness and fault tolerance determine the stable performance of the intelligent customer service system in the complex and changeable real user environment full of "noise." This requires the model to effectively address the challenges of diversity of user expressions, including colloquial expressions, omissions, typos, dialect influences, and even loose grammatical structures. At the same time, the system must demonstrate the ability to filter out distracting information, such as extraneous topics, emotional catharsis, or tentative questions mixed in the conversation. Evaluating this dimension often requires proactively designing stress tests to quantify the model's ability to maintain performance under non-ideal inputs by injecting noise samples or constructing adversarial test cases (e.g., deliberately creating ambiguity, inserting extraneous entities). This test can effectively reveal the boundaries and weaknesses of the model's comprehension capabilities.

Knowledge coverage is a basic metric to measure the scope of service system and the depth of support, which reflects the proportion of user questions that can be effectively answered by the system knowledge base or feature set. A system with insufficient coverage will frequently encounter out-of-domain (OOD) or have

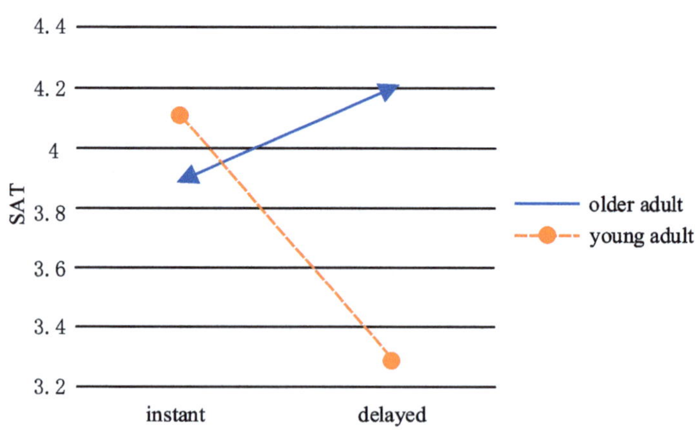

Fig. 3.4 Interaction of response time and age on SAT

to reject user requests, leading to reduced service availability and increased user frustration. Continuously monitoring OOD or request rejection rates, combined with user feedback to analyze high-frequency uncovered problem areas, is the core basis for driving knowledge base expansion and skill set optimization. High coverage is the basis for ensuring that users can obtain extensive service support through intelligent customer services channels, avoiding frequent manual transfers, and truly giving full play to the scale benefits of automated service system.

3.5.2 Advanced Evaluation Metrics

Multi-dimensional interaction experience evaluation has become the core direction of measuring intelligent customer services excellence, which goes beyond basic task completion requirements and focuses on the naturalness, empathy, and active interaction capabilities of human-computer dialogue. The naturalness evaluation metric is used to evaluate whether the content generated by the system conforms to human language habits, such as grammatical correctness, semantic coherence, and appropriateness. The automated evaluation framework based on the pre-trained language model can realize the quantification of this assessment, and the correlation between this framework and manual scoring significantly exceeds that of traditional text matching metrics [13]. Empathy ability needs to systematically identify the user's emotional state and generate responses with emotional support, which relies on the synergy of multimodal emotion recognition and emotion generation technology. Active interaction capabilities are used to evaluate context-aware prediction capabilities, which are reflected in advanced interaction behaviors such as active questioning, demand prediction, and suggestion generation, which mark a paradigm shift from "transaction processing" to "experience creation."

The Cognitive Depth metric assesses the ability of intelligent agents to handle complex scenarios. Factual accuracy is the lifeline of knowledge-intensive services, which requires strict alignment between external knowledge retrieval and generation, and in RAG architecture, it is necessary to strictly monitor the alignment of generated content with trusted knowledge sources [14]. Dialogue coherence requires maintaining the logical unity of information across rounds, including the consistency of entity referentiality, the stability of factual statements, and the continuity of arguments. Complex reasoning ability tests the multi-step problem-solving efficiency, involving causal inference, constraint integration, and operational sequence planning. Together, these metrics depict the cognitive boundaries of systematically dealing with unstructured problems.

3.6 Conclusion Remarks

The SE4SS framework systematically integrates scenarios engineering and new artificial intelligence technology to build a highly reliable and adaptable service paradigm for intelligent customer service systems. Its core lies in the strategic link between multi-level scenarios to open up the dynamic complexity of real scenes and the boundary expansion ability of generated scenes. Real-scenario data drives the accurate training of core components such as intent recognition and semantic analysis, while the generated scene breaks through data scarcity and ethical constraints through a controllable synthesis mechanism, rehearses extreme faults, and enhances system robustness. The construction and governance of high-quality datasets have become the fuel for intelligent evolution, and multi-dimensional annotation, privacy compliance, and bias correction technologies ensure that the data can not only reflect the complexity of real interactions, but also support the generalization needs of multi-task models. As the core cognitive engine, the large language model achieves a leap in service capability through multimodal fusion and neural symbolic reasoning. The modular architecture balances computing efficiency and performance requirements, while retrieval-augmented generation, explainable AI, and continuous learning mechanisms jointly build a reliable service closed loop. The multi-dimensional evaluation system forms panoramic monitoring from task success rate, cognitive depth to security governance, and promotes the evolution of the system from "transaction processing" to "experience creation."

The essence of SE4SS is to build a dynamic collaborative system of "scenario-data-model-evaluation": scenarios engineering defines service boundaries, data and models provide cognitive and execution capabilities, and the evaluation system guarantees the evolution direction. This framework upgrades customer service from passive response to user-centric active adaptation ecology and provides a reusable methodological basis for the engineering implementation of services science in the dimensions of coping with the chaos of the open world, meeting personalized needs and balancing efficiency responsibilities.

References

1. Robertson, R. E., Olteanu, A., Diaz, F., Shokouhi, M., & Bailey, P. (2021). 'I can't reply with that': Characterizing problematic email reply suggestions. In *Proceedings of the 2021 CHI Conference on Human Factors in Computing Systems (CHI '21, pp. 1–18)*. Association for Computing Machinery.
2. Liao, Y., Gao, Y., Wang, F., Zhang, L., Xu, Z., & Wu, Y. (2025). Emotion recognition with multiple physiological parameters based on ensemble learning. *Scientific Reports, 15*(1), 19869.
3. Warnat-Herresthal, S., et al. (2021). Swarm learning for decentralized and confidential clinical machine learning. *Nature, 594*(7862), 265–270.
4. Bran, A. M., et al. (2023). Augmenting large language models with chemistry tools. *Nature Machine Intelligence, 6*, 525–535.
5. Liu et al. (2021). A survey on dialogue systems: Recent advances and new frontiers. *IEEE Transactions on Knowledge and Data Engineering*.

6. Bai, X., Song, L., & Zhang, Y. (2022). Semantic-based pre-training for dialogue understanding. In *Proceedings of the 29th International Conference on Computational Linguistics* (pp. 592–607).
7. Jumper, J., Evans, R., Pritzel, A., et al. (2021). Highly accurate protein structure prediction with alphaFold. *Nature, 596*, 583–589.
8. Chowdhery, A., Narang, S., Devlin, J., et al. (2023). Palm: Scaling language modeling with pathways. *Journal of Machine Learning Research, 24*(240), 1–113.
9. Garcez, A. D. A., & Lamb, L. C. (2020). Neurosymbolic AI: The 3rd wave. *Artificial Intelligence Review* (pp. 1–20).
10. Rajbhandari, S., Li, C., Yao, Z. et al. (2022). Deepspeed-moe: Advancing mixture-of-experts inference and training to power next-generation AI scale. In *International conference on machine learning* (pp. 18332–18346).
11. Liu, C., Jiang, J., Xiong, C., Yang, Y., & Ye, J. (2022). Towards building an intelligent chatbot for customer service: Learning to respond at the appropriate time. In *Proceedings of the 26th ACM SIGKDD international conference on Knowledge Discovery & Data Mining* (pp. 3377–3385).
12. Wang, Y. L., & Lo, C. W. (2025). The effects of response time on older and young adults' interaction experience with chatbot. *BMC Psychol, 13*(1), 150.
13. Mehri, S., & Eskenazi, M. (2020). USR: An unsupervised and reference free evaluation metric for dialog generation. In *Proceedings of the 58th Annual Meeting of the Association for Computational Linguistics* (pp. 681–707).
14. Es, S., James, J., Anke, L. E., & Schockaert, S. (2024). Ragas: Automated evaluation of retrieval augmented generation. In *Proceedings of the 18th Conference of the European Chapter of the Association for Computational Linguistics: System Demonstrations* (pp. 150–158).

Chapter 4
SE4SS for Intelligent Hotels

4.1 Introduction

The hospitality industry is confronting major challenges in responding to the increasing demand for personalized, efficient, and scalable services. Traditional hotel management systems often struggle to meet the dynamic expectations of guests while ensuring operational efficiency. Issues such as inconsistent service quality, high labor costs, resource optimization, and data security continue to pose major barriers to progress. Additionally, the rapid growth of data and the need for real-time decision-making have made it increasingly difficult to maintain a competitive edge. As a result, the integration of artificial intelligence (AI) has become essential, yet ensuring that these AI-driven systems are reliable, transparent, and scalable remains a key concern [8, 12].

This chapter explores SE4SS as a solution to these challenges within the context of intelligent hotel. SE4SS combines advanced AI technologies with robust service engineering methodologies to enable hotel to automate routine operations, optimize resource management, and enhance the personalization of guest experiences. By integrating new AI into hotel systems, this framework ensures not only the efficiency and scalability of operations but also the reliability and ethical standards required for long-term success. Through SE4SS, intelligent hotel can enhance service delivery while maintaining accountability and security, addressing both operational needs and guest expectations.

4.2 Intelligent Hotel Robots

In the evolving hospitality industry, the demand for personalized, efficient, and scalable service system is increasing. One of the most promising innovations to meet these demands is the use of intelligent robots in hotel operations [13]. These robots,

X. Li and F.-Y. Wang, *Scenarios Engineering for Services Science*,
SpringerBriefs in Service Science, https://doi.org/10.1007/978-3-032-24152-8_4

powered by New AI, are revolutionizing the way hotel manages service delivery, from room service to concierge functions, all while ensuring operational efficiency and enhancing guest satisfaction [11]. This section explores how new AI enhances the capabilities of intelligent hotel robots, addressing the key challenges in the industry and transforming the service experience.

4.2.1 Complex Operational Scenarios

Intelligent hotel robots operate in complex environments where they must navigate not only physical space but also handle diverse, real-time service scenarios. These robots are tasked with performing multiple activities simultaneously, such as delivering items to rooms, assisting with check-ins, and interacting with guests in personalized ways. The complexity arises from the need to adapt to constantly changing conditions, such as fluctuating guest requests, varying occupancy rates, and unpredictable environmental factors (e.g., crowded hallways, unexpected obstacles). New AI enables robots to process real-time data from sensors and cameras, allowing them to continuously adapt and optimize their actions.

These complex scenarios can be further categorized into real-world scenarios (like guest room service) and generated scenarios (such as simulated environments for training robots). Generated scenarios are particularly useful for training robots to navigate unexpected situations and improve their response time. Real-world scenarios, on the other hand, provide the practical data that refine the AI models, ensuring that robots not only function in controlled environments but also perform well in dynamic hotel settings.

4.2.2 AI Agents and Autonomous Service Systems

A key component of intelligent hotel robots is the use of AI agents-autonomous systems capable of executing tasks without human intervention. These agents operate based on new AI models that integrate machine learning, natural language processing, and autonomous decision-making algorithms. AI agents enhance the hotel's ability to deliver autonomous service systems, such as autonomous check-in/check-out, room cleaning, and delivery services. These robots can engage with guests directly, offering service systems such as booking assistance, guiding guests to their rooms, or even answering questions regarding the hotel's amenities.

Autonomous service is particularly important in a fast-paced environment like a hotel, where customer expectations are high, and service speed is crucial. By leveraging New AI, robots are able to work autonomously in various hotel departments, improving efficiency, reducing human error, and enhancing the overall guest experience, as illustrated in Fig. 4.1.

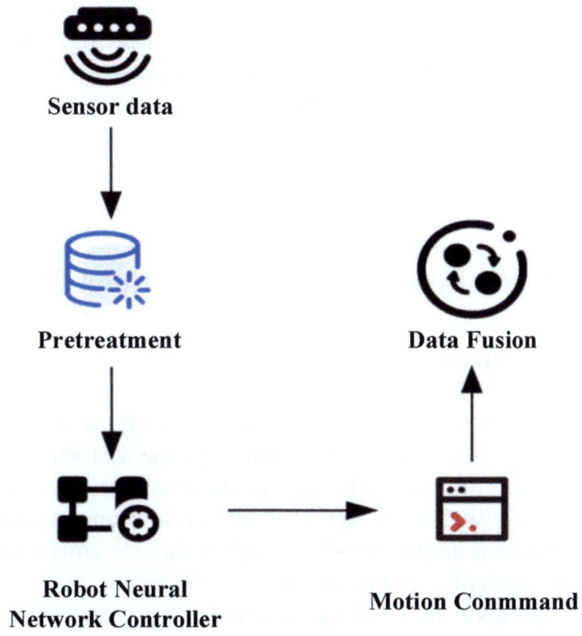

Fig. 4.1 Operation period of neural network motion controller

4.2.3 Large Models in Intelligent Hotel Robots

In intelligent hotel robots, large AI models are integral to enabling robots to operate autonomously and efficiently. These models are composed of multiple key components that process vast amounts of data and make real-time decisions. The neural networks (NN) used in these models allow robots to interpret sensory data, such as visual inputs from cameras, and navigate hotel environments. Convolutional Neural Networks (CNNs) are employed for obstacle detection and navigation, while Recurrent Neural Networks (RNNs) handle time-series data, such as service requests or guest interactions, improving task execution over time.

The integration of reinforcement learning (RL) within these models is critical for continuous optimization. RL enables robots to learn from their actions and adjust their behavior based on feedback, improving the efficiency of service delivery. By leveraging this approach, robots can autonomously refine their decision-making processes, selecting the best actions to take for tasks such as room service delivery or guest inquiries.

In addition to these techniques, natural language processing (NLP) models allow robots to interact with guests in a meaningful way, understanding and responding to requests in natural language. NLP helps bridge the gap between human-robot interactions, enabling robots to engage with guests on a personal level, whether it's answering questions or offering assistance.

These large models rely on continuous data input, which is essential for real-time decision-making and task optimization. Data from hotel sensors, guest interactions, and operational systems are constantly processed to improve robot performance and adapt to changing environments. The integration of knowledge automation ensures that robots can update their systems autonomously, leveraging centralized hotel data to enhance service delivery and remain aligned with the latest operational needs.

4.3 AI-Enhanced Sustainable Hotel Design

As global awareness of environmental issues grows and sustainable development goals (SDGs) become increasingly important, the hotel industry faces the dual challenge of reducing carbon emissions and improving energy efficiency [2]. However, while pursuing low-carbon buildings, the aesthetic design and guest satisfaction of hotel cannot be overlooked, as these factors directly impact the guest experience and loyalty. This is especially critical in economy chain hotel, which contributes a significant amount to global carbon dioxide emissions, approximately 300 million tons annually. Addressing this issue is urgent.

To tackle this challenge, the study [4] proposes an innovative solution: the integration of AI technologies to optimize prefabricated hotel renovations while coupling low-carbon strategies with aesthetic design. Specifically, the study employs multidimensional neural networks (NN), machine learning (ML), and statistical modeling (SM) to analyze and predict the impacts of energy efficiency, structural performance, and guest satisfaction during hotel renovations—methodological approaches that have been instrumental in advancing sustainable design research. SM is widely used to validate theoretical models and understand the relationships between latent variables [5], such as sustainability practices and design quality, while ML-NN models excel at handling complex, non-linear data patterns [15], providing high predictive accuracy. Additionally, 3D printing technology is used to manufacture prefabricated components, enhancing design flexibility and reducing material waste. This combined approach not only achieves low-carbon objectives but also maintains high-quality interior design and guest comfort.

The core of this solution lies in AI-driven renovations, which optimize energy consumption and structural efficiency while incorporating aesthetic design. This ensures that sustainable development goals are met without sacrificing guest satisfaction, ultimately achieving a balance between environmental sustainability and high-quality customer experience.

4.3.1 Sustainability Challenges

To effectively assess and optimize the renovation of prefabricated hotel buildings, this section investigates two distinct types of scenarios: real-world scenarios and synthetic scenarios. These scenarios provide a comprehensive framework for understanding how various renovation strategies can perform under different conditions.

Real-World Scenario: The real-world scenario focuses on the practical application of hotel renovations, incorporating actual data and conditions from existing prefabricated hotel. This includes evaluating energy consumption, space configurations, and guest satisfaction based on real-time feedback. Renovation processes are modeled based on these variables, which are often subject to environmental constraints, operational limitations, and varying guest preferences [3, 10]. By analyzing data from ongoing renovations, this scenario allows for the identification of potential challenges, such as balancing low-carbon goals with guest comfort. The real-world scenario provides insights into the complexities of applying sustainability measures while maintaining high standards of aesthetic quality.

Synthetic Scenario: In contrast, synthetic scenarios are based on computational simulations that model various renovation conditions, strategies, and technologies under controlled parameters. These scenarios are designed to test how different combinations of design elements, energy-saving technologies, and aesthetic choices influence energy efficiency, guest satisfaction, and carbon emissions. The use of synthetic scenarios enables the exploration of multiple renovation paths without the limitations imposed by existing buildings or operational constraints. This approach helps to identify the most promising strategies for achieving optimal outcomes, particularly in terms of sustainability and aesthetics.

By comparing results from both real-world and synthetic scenarios, this study provides a robust understanding of the potential impacts of different renovation strategies. The insights gained from these scenarios guide the optimization of renovation processes, ensuring that both low-carbon objectives and aesthetic standards are met effectively.

4.3.2 Datasets for Sustainable Hotel Design

For the purpose of training and evaluating ML and NN models, researchers utilize a variety of comprehensive datasets that are crucial for optimizing the renovation process of prefabricated hotel buildings. These datasets cover key factors such as energy efficiency, material selection, guest satisfaction, and carbon emissions, providing valuable input to enhance both sustainability and aesthetic design in the renovation process.

Energy Consumption Dataset: This dataset contains detailed records of energy usage in various parts of the hotel, including heating, ventilation, and air conditioning (HVAC) systems, lighting, and other electrical systems. By analyzing this data, the models can predict the potential improvements in energy efficiency resulting from different renovation strategies, such as upgrading insulation, replacing outdated systems with energy-efficient ones, or integrating renewable energy sources.

The dataset plays a crucial role in assessing how these strategies can help reduce the hotel's carbon footprint.

Design and Material Dataset: The design and material dataset includes information on the different construction materials and design elements used in the renovation process. It provides data on energy-efficient windows, sustainable flooring, and other eco-friendly building materials. This dataset is vital for helping the models optimize material selection, allowing for the identification of the most cost-effective, energy-efficient options that also meet aesthetic and functional requirements. Additionally, the dataset aids in evaluating the trade-offs between material costs, sustainability, and guest comfort.

Customer Satisfaction and Feedback Dataset: Guest satisfaction is an important consideration in hotel renovations. This dataset contains feedback collected from hotel guests regarding various aspects of the hotel's interior design, room layout, comfort, and overall aesthetics. This data is used to train the models to predict how renovation decisions will affect guest satisfaction and loyalty. The dataset also helps identify design features that are most valued by guests, such as space optimization, lighting quality, and the perceived environmental friendliness of the hotel.

Carbon Emissions Dataset: The carbon emissions dataset tracks the environmental impact of different renovation methods and materials. It provides insights into how various low-carbon strategies—such as the use of recycled materials, energy-efficient systems, and eco-friendly construction practices—contribute to reducing the hotel's carbon emissions. This dataset is essential for evaluating the sustainability of renovation solutions and ensuring that the hotel meets environmental standards while achieving aesthetic goals.

These datasets allow the models to integrate a wide range of variables, providing a comprehensive understanding of the renovation process. They also ensure that the predictions made by the models are grounded in real-world data, thereby improving the accuracy and relevance of the optimization outcomes.

4.3.3 Foundation Models for Hotel Renovation

In this section, foundation models are employed to handle the complexity and scale of data involved in the renovation process of prefabricated hotel buildings. These models are essential for processing large volumes of data and capturing intricate patterns that may not be apparent through traditional analysis methods. The use of large-scale ML and deep learning models enables the integration of multiple variables, optimizing both energy efficiency and aesthetic design.

The study utilizes traditional machine learning models, including regression analysis, clustering algorithms (such as K-Means), and principal component analysis (PCA). These models are effective for handling linear relationships between variables and identifying key factors that influence energy consumption, design choices,

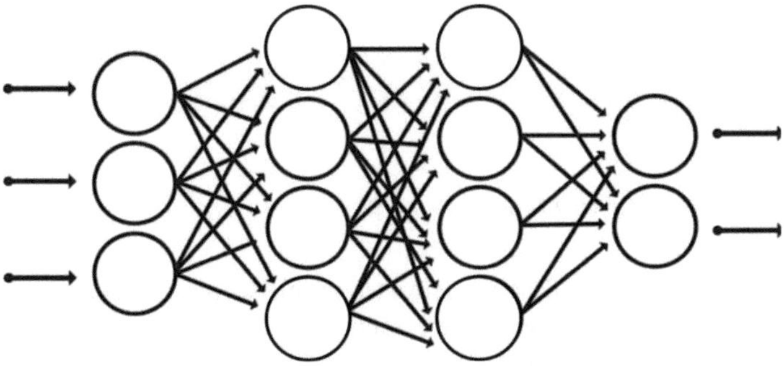

Fig. 4.2 Neural network

and guest satisfaction. Regression models, for example, predict the impact of specific renovation strategies on energy usage, while clustering algorithms help segment customers based on their preferences for design and sustainability features. PCA is employed to reduce the dimensionality of the data, simplifying complex datasets while retaining the most important features.

Additionally, foundation models, particularly NN, are used to handle non-linear relationships and complex datasets, as shown in Fig. 4.2. Neural networks, including multi-layer perceptrons (MLP) and convolutional neural networks (CNN), are capable of extracting high-dimensional features from the data that traditional machine learning methods might miss. These models are particularly valuable for identifying patterns in large-scale datasets, such as how different design choices influence both environmental performance and guest satisfaction. The capacity of foundation models to learn from vast amounts of data allows them to make highly accurate predictions regarding the outcomes of various renovation strategies.

One of the key advantages of using foundation models is their ability to integrate various types of data, such as energy consumption, material properties, customer feedback, and carbon emissions. The combined use of machine learning and neural network models enables a more nuanced understanding of the renovation process, taking into account the interdependencies between design elements, sustainability features, and guest preferences. This integration allows for the prediction of complex interactions and helps optimize renovation strategies for both aesthetic and environmental outcomes.

Given the size and complexity of the data, these foundation models require significant computational resources for training and optimization. High-performance computing infrastructure is utilized to run these models, ensuring the efficient processing of large datasets and the generation of real-time predictions. This allows the models to quickly adapt to new data and refine predictions as the renovation process progresses, making them highly adaptable to different scenarios and conditions.

By employing foundation models, this section provides a more accurate and holistic view of the renovation process, enabling a deeper understanding of the interactions between sustainability and aesthetic considerations. These models play a crucial role in optimizing the balance between energy efficiency, carbon reduction, and guest satisfaction, helping to create low-carbon, energy-efficient, and visually appealing hotel environments.

4.3.4 Evaluation Metrics

To assess the performance of the models used in this section, a variety of evaluation metrics are applied. These metrics are essential for determining the accuracy, reliability, and robustness of the models' predictions. They help identify model strengths, weaknesses, and guide further improvements.

- **R^2 (Coefficient of Determination)**: The R^2 metric is the primary measure used to evaluate how well the model explains the variance in the target variable. A higher R^2 value indicates that the model explains a larger proportion of the variability in the data. For example, an R^2 value of 0.78 suggests that 78% of the variance is explained by the model, which is considered a strong fit. This metric is particularly useful for regression models predicting continuous outcomes, such as energy consumption or guest satisfaction.
- **Mean Squared Error (MSE) and Root Mean Squared Error (RMSE)**: MSE and RMSE are used to measure the precision of the model's predictions. Both metrics calculate the average squared difference between the predicted and actual values. Smaller values indicate better model performance. RMSE, in particular, provides an interpretable value in the same units as the target variable, making it easier to understand the magnitude of the model's prediction errors.
- **Precision, Recall, and F1-Score**: These metrics are used for evaluating classification models, such as predicting guest satisfaction levels or categorizing renovation strategies based on their environmental impact. Precision measures the proportion of true positive predictions among all positive predictions made by the model. Recall measures the model's ability to identify all true positives (i.e., how well it detects positive cases). F1-Score is the harmonic mean of Precision and Recall, providing a balanced measure that accounts for both false positives and false negatives. It is particularly useful when dealing with imbalanced datasets.
- **Confusion Matrix**: The Confusion Matrix provides a detailed breakdown of classification performance by showing the number of true positives, false positives, true negatives, and false negatives. It is especially helpful for understanding how well the model classifies different categories, such as predicting which renovation strategies will most likely meet guest satisfaction or sustainability goals.
- **Goodness-of-Fit Indices**: In the context of Structural Equation Modeling (SEM), Goodness-of-Fit indices such as RMSEA (Root Mean Square Error of Approximation) and CFI (Comparative Fit Index) are used to assess how well the model fits the observed data. RMSEA values below 0.05 indicate a good fit between the model and the data, while CFI values above 0.90 suggest a well-fitting model. These indices help assess whether the proposed relationships between variables in the model are supported by the data.

By utilizing these evaluation metrics, the study ensures that the models are accurate, reliable, and capable of predicting outcomes under various renovation scenarios. These metrics help refine the models, ensuring that the proposed renovation strategies are effective and suitable for real-world applications.

4.4 Hotel Location Prediction

As urbanization accelerates and the tourism industry rapidly develops, hotel location selection has become increasingly complex and critical. A well-chosen hotel location not only enhances the market competitiveness of hotel but also contributes to optimizing urban spatial layouts and traffic systems. Star-rated hotel, as an important segment of the hotel industry, plays a significant role in determining tourism service quality and driving economic development. Therefore, accurately predicting the optimal locations for star-rated hotel is an urgent problem in both academia and industry [7].

Traditional hotel location research often relies on simple statistical analyses or geographic information systems (GIS) technologies [14]. However, these methods frequently face challenges such as limited data availability and lower model accuracy. In recent years, the application of AI and big data technologies has emerged as a promising solution to the hotel location problem [9], particularly when dealing with small datasets and complex environments. Specifically, meta-learning algorithms have shown significant potential in predicting hotel locations. By leveraging meta-learning models, it becomes possible to quickly adapt to different location prediction tasks with limited sample data, improving both accuracy and operational applicability.

Furthermore, traffic accessibility is a critical factor in hotel location selection that has gained increasing attention in recent studies [6]. Optimizing hotel location choices based on traffic convenience, road density, and the distribution of transportation facilities remains a key challenge in the modern hotel industry. In this context, this section proposes a method for hotel location prediction that integrates meta-learning and traffic accessibility analysis. The goal is to use data-driven intelligent decision-making to enable efficient and accurate hotel location selection in complex urban environments.

4.4.1 Meta-Learning

Meta-learning, also referred to as "learning to learn," has emerged as a powerful approach for addressing problems where data is sparse, or learning tasks differ slightly but share common underlying patterns. The core idea of meta-learning is to build a model that can generalize well from limited data by learning from previous learning experiences. This is particularly valuable in hotel location prediction, where data on successful hotel placements can be sparse, especially for newly developing areas.

In the context of hotel location prediction, meta-learning algorithms are applied to adapt to different regions, taking into account unique local factors such as socio-economic trends, transportation accessibility, and land usage. The model learns from past predictions, thereby improving its performance as it is exposed to new tasks with fewer data points. Through this approach, meta-learning allows the prediction of star-rated hotel locations with greater accuracy, even when historical data is limited.

A key component of meta-learning in this context is its ability to work with limited datasets, making it an ideal technique for locations where comprehensive historical data may not exist or where new hotel projects are being considered. By utilizing a few-shot learning paradigm, meta-learning models can leverage even a small number of training examples to make accurate predictions for unobserved regions or newly emerging locations. This is crucial in the rapidly changing urban environments where hotel locations must constantly evolve.

4.4.2 Neural Network Models

In this study, we integrate Long Short-Term Memory (LSTM) networks and K-means Random Forest (KMRF) within a meta-learning framework to develop an advanced solution for hotel location prediction, as illustrated in Fig. 4.3.

LSTM networks are used to address the temporal aspects of hotel site selection. Urban environments are dynamic, with factors like traffic patterns, economic conditions, and infrastructure development evolving over time. LSTMs excel at modeling these time-dependent factors by retaining long-term dependencies in the data. This ability allows the model to predict how these factors will change in the future, providing a robust foundation for forecasting hotel performance in emerging areas.

On the spatial side, the K-means Random Forest (KMRF) model helps address the complexity of hotel location features. The K-means clustering algorithm groups locations based on similar characteristics such as proximity to transport networks, population density, and socio-economic conditions. Random Forest is then applied to evaluate these clusters, identifying the most relevant factors that determine the success of a hotel in each region. This combination improves prediction accuracy, especially in areas with limited historical data, by focusing on clusters of similar locations.

The integration of these models within a meta-learning framework adds an additional layer of adaptability. Meta-learning enables the system to learn from previous tasks and apply this knowledge to new ones, even when faced with limited data. This enhances the model's ability to generalize and predict hotel locations in regions where data may be sparse or unavailable.

Together, these neural network models form a powerful, adaptable solution for predicting the best locations for star-rated hotel. By considering both temporal and

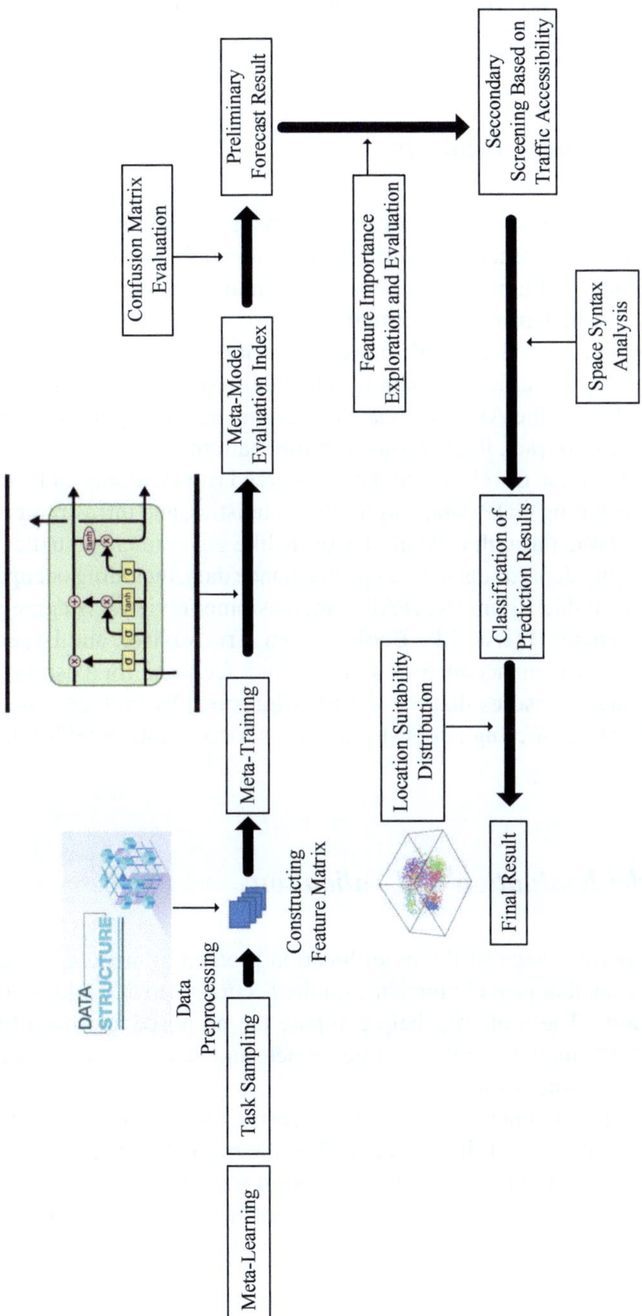

Fig. 4.3 Research frame diagram

spatial factors, and leveraging meta-learning to improve adaptability, this approach offers accurate, scalable predictions that can handle the complexities of modern urban environments.

4.4.3 Multi-Source Datasets

The dataset for hotel location prediction is sourced from multiple reliable and diverse datasets to ensure comprehensive coverage of factors influencing hotel success. Spatial data is primarily obtained from geospatial databases and mapping services like OpenStreetMap, which provide information on proximity to transportation hubs (airports, train stations), road network density, and land use types (residential, commercial, mixed-use). Traffic flow data is collected from real-time traffic monitoring systems and public traffic APIs, such as those from local transportation departments and Google Maps, to track fluctuations in traffic patterns.

Economic and urban development data is sourced from national and local government reports, including GDP data, employment statistics, and infrastructure development plans available through official platforms like government statistical agencies and urban planning departments. Hotel performance data, including occupancy rates, Revenue Per Available Room (RevPAR), and customer reviews, is aggregated from online travel agencies (OTAs) like Booking.com, TripAdvisor, and Expedia, which provide detailed performance metrics and customer feedback for existing hotel. This multi-source dataset ensures that the model is informed by both current conditions and historical trends, offering a robust foundation for accurate hotel location predictions.

4.4.4 Model Evaluation and Validation

To assess the performance of the hotel location prediction models, we use several evaluation metrics that provide insight into the accuracy, error, and generalizability of the predictions. These metrics help compare the proposed approach (integrating LSTM and KMRF models) with baseline models and determine its effectiveness in predicting optimal hotel locations.

Accuracy is a fundamental metric that measures the proportion of correct predictions made by the model. It is calculated as the ratio of correctly predicted hotel locations to the total number of predictions. High accuracy indicates that the model is effectively predicting the success or failure of hotel locations, especially when considering factors like accessibility, demand, and socio-economic conditions.

$$\text{Accuracy} = \frac{\text{Number of Correct Predictions}}{\text{Total Number of Predictions}}$$

Mean Absolute Error (MAE) quantifies the average magnitude of errors in the model's predictions. It calculates the average of the absolute differences between predicted and actual values, providing a straightforward measure of the model's prediction accuracy. Lower MAE values indicate better predictive performance. MAE is particularly useful for continuous variables such as occupancy rates or revenue, where the model's goal is to make fine-grained, precise predictions.
where

- y_i is the true value.
- $\hat{y}_i$ is the predicted value.
- n is the total number of data points.

Root Mean Squared Error (RMSE) is another widely used metric that measures the average magnitude of errors, but with higher penalties for larger errors. RMSE is particularly useful when large errors are considered more detrimental to the model's performance. It is calculated by taking the square root of the average squared differences between predicted and actual values. Lower RMSE values indicate that the model's predictions are closer to the true values.

$$\text{RMSE} = \sqrt{\frac{1}{n}\sum_{i=1}^{n}(y_i - \hat{y}_i)^2}$$

F1-score is a metric used for evaluating classification models, particularly in imbalanced datasets. It is the harmonic mean of precision and recall, offering a balanced view of the model's ability to correctly classify successful versus unsuccessful hotel locations. The F1-score is useful when the dataset has more successful or unsuccessful locations, ensuring that the model performs well across both classes without being biased toward the majority class.

$$\text{F1-Score} = 2 \times \frac{\text{Precision} \times \text{Recall}}{\text{Precision} + \text{Recall}}$$

where

- **Precision** is the proportion of true positives among all positive predictions.
- **Recall** is the proportion of true positives among all actual positives.

AUC-ROC curve provides a graphical representation of the model's classification performance. The Area Under the Curve (AUC) measures the model's ability to distinguish between classes, where a higher AUC indicates better discrimination. A model with an AUC of 1 is perfect, while an AUC of 0.5 suggests no discriminative ability.

$$\text{AUC} = \int_0^1 \text{True Positive Rate}(x)\, d\text{False Positive Rate}(x)$$

To assess the model's robustness and generalizability, the study performs **k-fold cross-validation**. This involves splitting the dataset into k subsets, training the model on $k - 1$ subsets, and testing it on the remaining subset. The process is repeated for each subset, and the final evaluation metric is averaged over all folds. This method ensures that the model performs consistently across different subsets of data, providing a more reliable measure of its predictive power.

Experimental Results and Discussion

In this section, we present the results of our experiments evaluating the proposed hotel location prediction model, which integrates LSTM (Long Short-Term Memory) and K-means Random Forest (KMRF) within a meta-learning framework. We assess the model's performance based on various evaluation metrics and compare it with baseline models to determine its effectiveness in predicting optimal hotel locations.

We first evaluated the proposed model using a test dataset, comparing its performance to baseline models, such as traditional statistical methods and simpler machine learning algorithms like Decision Trees and Support Vector Machines (SVM). The results show that the integrated LSTM-KMRF model significantly outperforms these baseline models across several key performance metrics.

Accuracy: The LSTM-KMRF model achieved an impressive accuracy of 90.45%, while the Decision Tree and SVM models only reached accuracies of 78.30% and 81.50%, respectively.

Mean Absolute Error (MAE): The proposed model achieved a MAE of 0.13, which is considerably lower than the 0.19 achieved by the Decision Tree model and the 0.18 by SVM.

Root Mean Squared Error (RMSE): The RMSE for the LSTM-KMRF model was 0.22, which again showed superior performance compared to the Decision Tree (RMSE = 0.30) and SVM (RMSE = 0.29).

These results indicate that our model not only provides more accurate predictions but also reduces prediction errors, making it a more reliable solution for hotel location prediction.

We also tested the model's ability to generalize to new data by performing k-fold cross-validation across different regions with varying amounts of data. The LSTM-KMRF model consistently outperformed the baseline models, even when trained on smaller datasets. This demonstrates the strength of the meta-learning approach, allowing the model to quickly adapt and make accurate predictions even when only limited data is available for new regions.

The combination of LSTM for temporal analysis and KMRF for spatial clustering proves to be highly effective for predicting hotel locations. By accounting for dynamic factors like traffic flow, economic trends, and urban development over time, as well as spatial features like proximity to transportation hubs and population density, our model delivers a more comprehensive and accurate prediction compared to traditional approaches.

In conclusion, the LSTM-KMRF model integrated with meta-learning offers a robust, scalable, and highly accurate solution for predicting hotel locations. It outperforms traditional models, demonstrating not only improved predictive accuracy but also strong generalization capabilities across diverse datasets and regions.

4.5 Conclusion Remarks

This chapter applies the SE4SS framework to the development of intelligent hotel, illustrating how new AI technologies—such as embodied intelligence, large language models, reinforcement learning, and meta-learning—can transform hospitality service system [1]. By employing intelligent robots, hotels can deliver autonomous and efficient service systems that improve guest experiences while lowering operational costs. Furthermore, AI-driven sustainable design demonstrates how prefabricated hotel renovations can simultaneously optimize energy efficiency, structural performance, and aesthetic quality. Foundation models and comprehensive datasets enable data-informed renovation strategies that balance sustainability and customer satisfaction. The chapter also introduced a meta-learning-based approach to hotel location prediction, integrating LSTM and KMRF models to accurately forecast optimal sites under complex spatialâŁ"temporal dynamics. Overall, SE4SS provides a systematic, transparent, and trustworthy pathway for integrating AI into hotel operations, ensuring efficiency, personalization, and ethical accountability in the next generation of intelligent hospitality systems.

References

1. Alawami, A., Alawami, M., Obaid, A., Al-rushaydan, M., Boudjedra, M. L., Bou-djedra, B. E., Alharbi, M., Alzoori, H., Ab-osaif, S., Bshir, G., Alawami, H., & Abufawr, M. (2025). A systematic review of AI-driven innovations in the hospitality sector: Implications on restaurant management. *American Journal of Industrial and Business Management,15*, 30–40.
2. Cai, G., Sun, L., Liu, D., Xu, B., & Mo, Z. (2025). Potential of indoor room 3D ratio in reducing carbon emissions by prefabricated decoration in chain hotel buildings via multidimensional algorithm models for robot in-situ 3D printing. *Journal Of Building Engineering,101*, Article 111757.
3. Cai, G., et al. (2022). Knowledge archaeology on relations between the Venice architecture biennale (1980–2018) and the Pritzker architecture prize (1979–2019). *Journal Of Asian Architecture And Building Engineering, 21*, 224–233.
4. Cai, G., Lou, Y., & Lu, F. (2025). AI enhancing prefabricated aesthetics and low carbon coupled with 3D printing in chain hotel buildings from multidimensional neural networks. *Scientific Reports, 15*, 13229.
5. Chi, X., Zhou, H., Cai, G., & Han, H. (2024). Investigation into the festival brand co-creation mechanism: Extended application of the customer-based brand equity model. *Journal of Travel & Tourism Marketing, 41*, 377–395.

6. Huang, Q., Yang, B., Xu, X., Hao, H., Liang, L., & Wang, M. (2022). Location selection and prediction of SexyTea store in Changsha City based on multi-source spatial data and random forest model. *Journal of Geo-Information Science, 24*(4), 723–737.
7. Lai, C. Q., & Wu, X. B. (2019). A study on spatial distribution characteristics of megacity hotels based on GIS method: A data analysis based on POI data of Guangzhou hotels. *Modern Urban Research, 8*, 66–74.
8. Limna, P. (2022). Artificial Intelligence (AI) in the hospitality industry: A review article. *International Journal of Computing Sciences Research, 6*, 1–12.
9. Li, N., & Wu, H. (2025). Few-shot hotel industry site selection prediction method based on meta learning algorithms and transportation accessibility. *Scientific Reports, 15*, 16109.
10. Wise, N., & Mulec, I. (2015). Aesthetic awareness and spectacle: Communicated images of Novi sad (Serbia), the exit festival, and the Petrovaradin fortress. *Tourism Review International, 19*, 193–205.
11. Xu, J., Hsiao, A., Reid, S., & Ma, E. (2023). Working with service robots? A systematic literature review of hospitality employees' perspectives. *International Journal of Hospitality Management,113*, Article 103523.
12. Zahidi, F., Kaluvilla, B. B., & Mulla, T. (2024). Embracing the new era: Artificial intelligence and its multifaceted impact on the hospitality industry. *Journal of Open Innovation: Technology, Market, and Complexity,10*(4), Article 100390.
13. Zhang, J. (2025). Application and development of artificial intelligence and computer technology in the field of intelligent robots. *Scientific Reports, 15*, 14656.
14. Zha, A. P., Xu, N., & Hou, Z. G. (2017). Research on the suitability of micro-location of budget hotel: A case study on Jinjiang Inn in Shanghai Central City. *Human Geography, 32*, 152–160.
15. Zhou, Z., Qiu, C., & Zhang, Y. (2023). A comparative analysis of linear regression, neural networks and random forest regression for predicting air ozone employing soft sensor models. *Scientific Reports, 13*, 22420.

Chapter 5
SE4SS for Intelligent Cities

5.1 Introduction

Contemporary intelligent city development is undergoing a paradigm shift from a mere stacking of technologies to ecological synergy. Its essence is to restructure the "perception-decision-execution" closed loop in urban governance through digital technologies, aiming to achieve the triple objectives of enhanced efficiency, sustainable development, and improved resident well-being. The DTUMOS (Digital Twin for Urban Mobility Operating System) proposed by Hyeokju Yeon et al. [1] in 2023 validates this approach via a digital twin framework, as shown in Fig. 5.1. By employing an AI-driven ETA model and vehicle routing algorithms, DTUMOS achieved a simulation speed more than twice as fast as existing digital twin systems in field tests conducted in major metropolitan areas like Seoul and New York. Its four-module architecture (Data, AI Models, Operating System, Output) has not only significantly improved the efficiency of traffic policy rehearsals but also, through case studies, demonstrated the value of digital twins in practical urban governance.

The urban digital twin technology, exemplified by DTUMOS, holds profound value that extends far beyond traffic flow optimization. It signifies a paradigm shift in intelligent city research from an isolated, technology-driven approach to a systemic "Eco-Digital Integration." This new paradigm aims to build an urban governance framework characterized by deep synergy between digital technologies and natural systems, thereby systematically addressing the complex trade-offs among environmental, economic, and social well-being. In the realm of high-precision ecological perception, the integration of AI and remote sensing is deciphering the "urban ecological code" in unprecedented ways. For example, models based on deep Convolutional Neural Networks (CNNs) can automatically and accurately segment green spaces, water bodies, and impervious surfaces from high-resolution satellite imagery—a process that not only enables near-daily monitoring of dynamic changes in urban green cover but also facilitates the quantification of associated ecosystem service values, such as carbon sequestration, oxygen release, cooling, and humidity enhancement [2].

X. Li and F.-Y. Wang, *Scenarios Engineering for Services Science*,
SpringerBriefs in Service Science, https://doi.org/10.1007/978-3-032-24152-8_5

Going a step further, emerging technologies can even assess and map habitat quality for urban biodiversity by integrating multi-source remote sensing data (e.g., hyperspectral, thermal infrared, and LiDAR) with AI algorithms [3], thereby providing a scientific basis for targeted interventions in biodiversity conservation, as illustrated in Fig. 5.2.

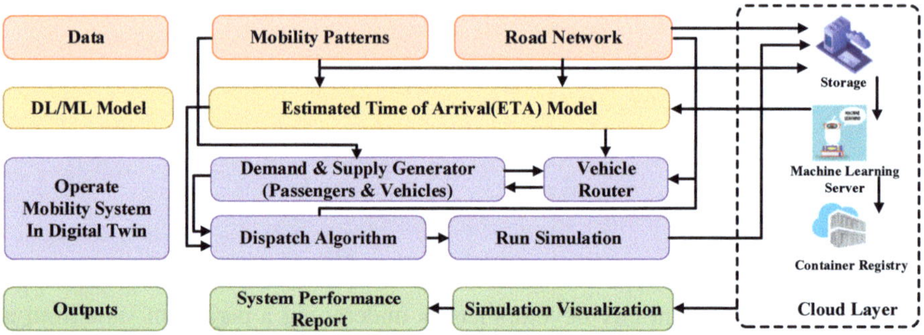

Fig. 5.1 Illustration of the framework of DTUMOS

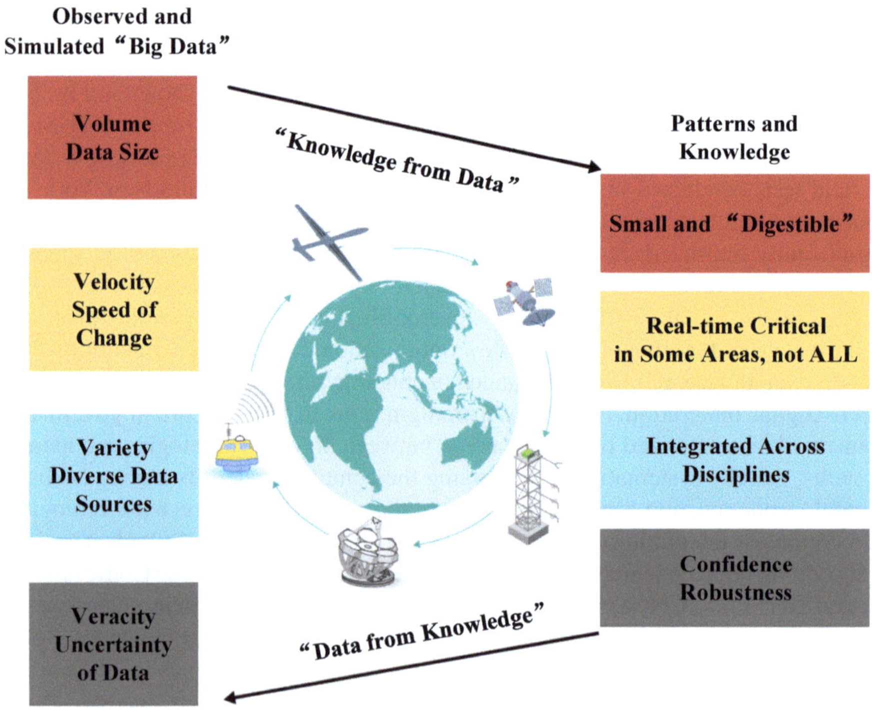

Fig. 5.2 Big data challenges in the geoscientific context

At the level of systemic optimization, intelligent city technologies are indispensable core infrastructure for achieving deep urban decarbonization. Their emission reduction potential is rooted in the capacity to perform nonlinear, synergistic optimization of complex giant systems like urban energy, transportation, and building networks. Relevant research indicates that, compared to merely improving energy supply-side efficiency, digital technology-based demand-side management—such as adaptive smart grids and on-demand responsive shared mobility—represents a more cost-effective leverage point with greater emission reduction potential [4]. In this context, urban digital twins serve as critical "policy simulators." They can create a virtual urban system to run large-scale, low-cost what-if simulations for various planning policies (e.g., adding bus lanes, raising parking fees, promoting rooftop photovoltaics). This enables the quantitative prediction of each scenario's carbon reduction effects, economic costs, and social impacts, thereby providing empirical evidence to help policymakers identify the most cost-effective path [5]. The success of intelligent cities therefore depends not only on breakthroughs in individual technologies but, more crucially, on the systematic integration of digital technologies into the ecological logic of urban governance, ultimately achieving synergistic progress in environmental, economic, and social development. The sustainable development of intelligent city relies on the continuous, low-cost monitoring of urban surface environments, including green spaces, roads, and building density. The MOSAIKS method proposed by Esther Rolf et al. [6] is a key breakthrough in this regard. It uses satellite imagery and randomly generated convolutional features to create a universal sensing framework based on "compute once, reuse for many tasks." This allows urban managers to quickly obtain predictions for key ecological indicators like forest cover and road density, significantly lowering the barrier to deploying intelligent city data infrastructure. Although the method has limited predictive power for socioeconomic variables (e.g., income, housing prices), it provides an efficient, scalable, and democratized monitoring solution for physical environmental features visible in satellite images, laying a data foundation for ecological assessment and spatial planning in intelligent city.

In addition to ecological and efficiency benefits, intelligent city architectures are emerging as a core pillar for building resilient cities by enhancing public health emergency response capabilities. Their transformative potential originates from the ability to shift the traditional passive and lagging emergency management model toward a new paradigm characterized by big data-driven early warning, model-based precise prediction, and simulation-informed dynamic intervention [7]. The technical foundation of this transformation lies in the integration and intelligent analysis of urban multi-source heterogeneous data. For instance, by leveraging anonymized mobile device location data (e.g., CDR, GPS), these systems [7] can generate high-resolution population contact networks, providing crucial fine-grained clustering data inputs for infectious disease dynamics models.

The intelligent city platform can function as a powerful simulation tool. As demonstrated by multi-layer network models constructed from real urban data, it allows decision-makers to conduct large-scale, repeated stochastic simulations of various public health interventions—such as adjusting testing rates or scopes of contact tracing. These simulations yield crucial quantitative metrics, including ICU bed demand and peak infection rates. This ex ante quantitative assessment provides robust evidence for selecting optimal policy pathways and facilitates the analysis of trade-offs between public health and economic activity. Global modeling studies indicate that integrating mobility data with proactive intervention strategies, such as contact tracing, can effectively flatten [8] the epidemic curve and prevent the overload of critical healthcare resources. Furthermore, by leveraging a dynamically updated contact network that changes daily, the model can estimate future hospitalization needs in near real time, thereby offering a critical evidence base for the proactive allocation of medical resources.

The core breakthrough in contemporary intelligent city hinges on dual advancements in "data fluidity" and "decision-making autonomy." This entails the deep integration of disparate sensing devices (e.g., RSUs, environmental sensors), heterogeneous data sources (e.g., government records, trajectory data, remote sensing imagery), intelligent decision-making models (e.g., Multi-Agent Systems, large language models), and urban infrastructure (e.g., road networks, power grids, utility pipelines). According to a 2025 study by Kan Wu et al. [9], which utilized city-level traffic simulation and emission model evaluation, the large-scale deployment of such systems—despite their significant environmental and economic benefits—remains contingent on high-quality, real-time data support. Many cities still face challenges related to inadequate sensing infrastructure and limited data integration capabilities. Addressing these gaps urgently requires the adoption of lightweight models, low-cost sensors, and public-private data collaboration mechanisms to foster the inclusive and sustainable deployment of intelligent city technologies.

5.2 Technical Architecture for Intelligent Cities

5.2.1 Multimodal Dataset Systems

The core mission of the intelligent city perception layer is to construct a "full-scenario, high-fidelity" digital representation of the urban environment. This foundation requires breaking down data silos and achieving the synergistic integration of multimodal data. A typical framework for this purpose integrates four primary data categories: (1) structured data (e.g., traffic flow, administrative records); (2) unstructured data (e.g., remote sensing imagery, street-view videos); (3) dynamic sensor data (e.g., environmental monitoring readings, equipment operational status); and (4) cross-departmental government data (e.g., inter-agency approval information). Current frontier research in this domain focuses on addressing the challenges of multi-source heterogeneous data fusion. Adherence to the FAIR principles—making data Findable, Accessible, Interoperable, and Reusable—has been established as the cornerstone of scientific data sharing and integration, serving as a prerequisite for building such complex data ecosystems [10]. Specifically within the intelligent city context, the integration of geospatial data (e.g., remote sensing images, sensor locations) heavily relies on international standards like those from the Open Geospatial Consortium (OGC). These standards ensure that data from diverse sources can be accurately overlaid, analyzed, and visualized within a unified spatial framework. Concurrently, embedding the "Privacy by Design" philosophy at the outset of the data life cycle, along with establishing a cautious data governance and ethical framework, is critical for balancing technological innovation with the protection of citizens' rights.

5.2.2 Environmental and Traffic Datasets

High-precision environmental monitoring depends on the deployment of sensor networks and rigorous data quality control, featuring a synergistic "fixed + mobile" architecture. The validity of data [11] from these mobile units hinges on real-time calibration algorithms and robust data transmission links. The introduction of edge computing has substantially improved data processing efficiency. Traffic data acquisition necessitates the development of a multi-dimensional behavioral profile system. By leveraging and cross-validating multi-source data, such as GPS trajectories and IC card transaction records, it becomes possible to reconstruct the movement patterns of pedestrians and vehicles.

5.2.3 Government Affairs and Residents' Behavior Data

Government data integration is pivotal for enhancing public service efficiency. Its core principle involves establishing a cross-departmental data collaboration mechanism, favoring federation over physical centralization. By implementing a federated data platform, disparate departments can perform collaborative computations while keeping data localized. This model effectively breaks down "Data Silos" while ensuring security through robust permission management and encryption technologies. It thereby enables the correlated analysis of data from sectors like environment, transport, and health without compromising sensitive information. The digital transformation of government processes provides a natural vehicle for this integration. When procedures such as business registration and real estate transactions are fully digitized, the data generated during approval automatically flows into standardized databases. Semantic alignment is achieved through a unified data dictionary, which maps fields like "Enterprise Address" and "Legal Representative Information" from various departments onto common standards, laying the groundwork for cross-domain analysis. This integration directly translates into enhanced service efficiency, typically evidenced by streamlined approval procedures, reduced processing times, and more precise policy formulation. For example, analyzing the spatial distribution [12] of enterprise registration data enables more scientific planning of industrial park layouts.

5.3 New AI-Driven Intelligent City Decision Systems

The decision-making requirements in intelligent city are characterized by "distributed execution, dynamic variability, and multi-objective balancing." These requirements span from resolving local, real-time issues like traffic signal control and energy dispatch to coordinating global, long-term objectives such as urban ecology and public services. The inherent complexity of these tasks precludes the use of any single algorithmic solution. Consequently, a deep integration of Multi-Agent Systems (MAS) with advanced artificial intelligence has emerged as the core strategy. Within this framework, MAS is tasked with decomposing decision-making problems and enabling distributed coordination, while Large Language Models (LLMs) and Reinforcement Learning (RL) enhance "cognitive capacity" for complex problem-solving and "adaptive capability" in dynamic environments, respectively. Together, these technologies form a closed-loop decision chain described as "task decomposition → strategy generation → dynamic adjustment."

5.3.1 Multi-Agent Systems and Federated Learning

The core value of a multi-agent system lies in its ability to decompose complex urban decision-making into multiple executable local tasks, with each task managed by a dedicated agent. In the transportation domain, for instance, signal control agents at individual intersections, traffic flow coordination agents for regional road networks, and public transit scheduling agents collectively form an intelligent decision-making network. However, traditional MAS architectures are susceptible to "single-point optimization conflicts." A typical scenario occurs when an intersection agent, aiming solely to maximize its own traffic throughput by extending green lights, inadvertently causes congestion in adjacent road sections. Federated learning [13] addresses this limitation by enabling a "data non-sharing but model co-optimization" synergy, allowing agents to collaboratively improve their strategies without exchanging raw, potentially sensitive data.

The application logic of federated learning within MAS is clear. Each agent trains a local model based on its own data and uploads only the model parameters—not the raw data—to a coordination center. The center aggregates and optimizes these parameters before distributing them back to the agents for local model updating. This approach is uniquely suited to the reality of intelligent city, where data is dispersed across departments and difficult to centralize. It also mitigates the limitations of localized decision-making. This collaboration relies not on complex data pipelines but on the core technical principle of "parameter sharing instead of data sharing," making it highly practical for real-world deployment.

5.3.2 LLMs in Urban Decision-Making

LLMs offer intelligent city decision-making the capability to "process complex information and generate executable strategies." However, general-purpose LLMs cannot be directly applied to urban contexts. Urban decision-making must integrate physical constraints (e.g., road network topology is not malleable), policy requirements (e.g., prioritizing public services), and humanistic needs (e.g., travel convenience for the elderly). Consequently, it is [14] essential to develop city-specific foundation models through "domain knowledge injection," as illustrated in Fig. 5.3.

5.3.3 Multimodel Collaboration

Multimodel collaboration in intelligent city is not a simple superposition but rather forms a closed loop of "MAS for execution, LLMs for planning, and RL for adjustment." This loop operates as follows: first, LLMs generate an overall strategy based on the city's global objectives; second, the MAS decomposes this strategy into local tasks for individual agents; finally, RL adjusts the agents' execution plans based on real-time environmental feedback to prevent deviation from the overall goals. To ensure the reliability of this collaboration, a "decision validation mechanism" must be established. One aspect involves using the urban digital twin to simulate the effects of decisions. For example, traffic strategies generated by LLMs can be input into the twin system to simulate 12-hour traffic flow evolution. If potential congestion is predicted in a specific road section, RL can be proactively invoked to optimize the strategy. Another aspect involves incorporating human oversight, where

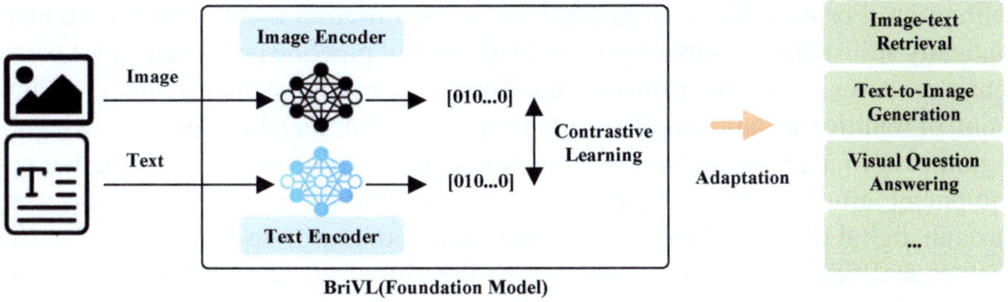

Fig. 5.3 Overarching concept of our BriVL model with weak training data assumption

high-risk decisions generated by the models—such as large-scale peak load shifting in the power grid or traffic control for major events—are subject to manual review to mitigate technical biases.

5.4 Mapping for Urban Digital Twins

Digital twin technology has evolved from single-point applications in transportation to diverse domains such as energy, building management, and public health, forming a comprehensive digital twin ecosystem that encompasses all urban elements. Its core breakthrough lies in establishing a real-time, consistent mapping between the virtual system and the physical world through high-precision modeling and dynamic error-correction techniques. This mapping is not a simple static replica but, via continuous data interaction and model refinement, enables the digital mirror to accurately reflect the dynamic state of physical entities [15]. The critical challenge in this evolution—from isolated, single-scenario simulations to a holistic, city-scale twin—is resolving fundamental issues like multi-source data fusion, cross-domain model interoperability, and virtual-real closed-loop feedback. The ultimate goal is to achieve end-to-end intelligence across the entire "perception-simulation-decision-execution" chain.

5.4.1 The Technical Architecture of All-Domain Twins

The technical architecture of a city-scale digital twin typically follows a three-tiered "Data-Model-Interaction" framework. The Data Layer is responsible for integrating multimodal heterogeneous data. This involves capturing the real-time status of the physical city via IoT sensors, establishing a static spatial baseline using Building Information Modeling (BIM) and Geographic Information Systems (GIS), and incorporating structured data (e.g., population, economic metrics) from government systems. The Model Layer centers on dynamic modeling and an error-correction mechanism. A synergy of physical rules and AI algorithms enables model self-optimization. First, by embedding "process mechanics," the model's physical plausibility is guaranteed. Second, real-time data is leveraged for continuous parameter calibration. For example, key parameters like the friction coefficient can be automatically adjusted by comparing virtually predicted pipeline pressure against actual sensor readings. This "mechanism + data" dual-driven approach overcomes the limitation of traditional static models in adapting to environmental changes, allowing the digital twin to accurately simulate complex scenarios such as extreme weather and equipment failures. The Interaction Layer achieves life cycle-wide data connectivity through digital thread technology. In construction, digital-twin-assisted engineering enables real-time synchronization between the virtual model and the physical site. Positional changes of construction machinery, captured by positioning systems, are fed back to the model, while the model's process optimization instructions simultaneously guide on-site operations, creating a closed loop for bidirectional dynamic adjustment. This capability breaks down information silos between the design, construction, and operation/maintenance phases, ensuring that decisions at every stage are informed by a comprehensive digital mirror.

5.4.2 Cross-Domain Applications of Digital Cities

Digital twins in the energy sector have enabled synergistic optimization across multiple systems. An urban energy digital twin platform integrates subsystems including the power grid, heating networks, and building energy consumption, using virtual simulation to evaluate energy dispatch schemes under various scenarios. In public health, digital twins focus on simulating transmission dynamics and intervention effectiveness. By constructing a virtual city model incorporating population mobility, environmental parameters, and healthcare resources, it becomes possible to simulate epidemic spread pathways and assess the impact of containment measures. This design allows for effective cost-benefit analysis of interventions, such as adjustments to lockdown areas, thereby providing a scientific basis for emergency decision-making [16]. The convergence of digital twins for buildings and transportation enhances urban spatial efficiency. In high-rise construction, digital twins can simulate scenarios like structural loads and equipment scheduling, identifying potential conflicts beforehand. Post-commissioning, operational strategies are continuously refined by collecting real-time data on energy use and equipment status. Within the transportation sector, simulation has evolved from individual road segments to a twin of the entire network. This system not only predicts traffic flow variations but also interfaces with signal control, public transit dispatch, and other systems. By identifying the optimal solution through virtual commissioning before real-world implementation, it significantly reduces the costs associated with trial and error on physical road networks.

5.4.3 Virtual-Physical Interaction in Urban Scenarios

The essence of virtual-real interaction lies in establishing a dynamic closed loop of "perception-simulation-decision-feedback." Data on the state of the physical world is fed in real time to the virtual model via sensors. The model then runs multi-scenario simulations to generate optimized strategies, which are deployed in reality by execution systems. The results of this implementation are subsequently fed back into the model, completing an iterative cycle. In traffic signal control, this loop operates as follows: intersection cameras collect traffic flow data; the digital twin simulates the traffic efficiency of different signal timing plans; the optimal plan is selected and dispatched to the signal controller; and finally, the actual traffic passage time is fed back to the model for parameter calibration.

The reliability of a digital twin depends on multiple safeguarding techniques. For data quality, outliers are eliminated through multi-source validation. For example, traffic camera data can be cross-validated with inductive loop detector data to ensure the accuracy of the foundational information input into the model. Model verification employs a "digital sandbox" approach, where historical events—such as past cold waves or epidemics—are replicated in the virtual environment. The model is only deployed to guide decision-making if its predictions demonstrate a strong correlation with the actual historical outcomes.

5.5 Urban Energy and Environmental Governance

Urban energy and environmental governance is undergoing a paradigm shift from "reactive remediation" to "proactive source regulation." In the traditional model, energy supply relies on emergency capacity expansion during peak demand periods, while environmental management typically involves passive control only after pollution exceeds standards. This reactive approach not only increases costs but also fails to address root causes. Modern intelligent governance systems enable proactive optimization through technological integration: building energy systems utilize spatiotemporal scheduling to shave peak loads; microgrids balance supply and demand by integrating distributed energy resources; mobile monitoring technologies accurately pinpoint pollution sources; and carbon accounting frameworks provide a scientific basis for emission reduction policies. These four technical pathways mutually reinforce each other, forming an end-to-end governance cycle of "prediction, regulation, monitoring, and optimization" that drives the transition from extensive to precision management in urban ecological governance.

5.5.1 Building Energy Optimization

Buildings, being the core sector of urban energy consumption, face the critical challenge of overcoming "concentrated peak demand and inefficient energy use." In hot and humid regions, air-conditioning loads can constitute over 0.6 of a building's total energy consumption. Traditionally, these systems often run at full capacity during peak hours, increasing grid stress and operational costs. The combined application of pre-cooling and virtual storage technologies offers a viable solution. By storing cooling energy during off-peak hours, this approach achieves spatiotemporal shifting of energy consumption. Virtual energy storage leverages a building's intrinsic thermal inertia for energy regulation. Structural components like walls and furniture in residential or commercial buildings possess inherent thermal mass. Using model predictive control (MPC) algorithms, air-conditioning systems can perform pre-cooling during low-tariff periods, lowering the indoor temperature by 2–3 °C. During peak hours, partial equipment can be shut down, maintaining comfort by utilizing the stored "coolth." The core principle of this strategy is analogous to "using the building as a battery," enabling flexible energy dispatch through thermal variation rather than chemical storage [17].

5.5.2 *Urban Microgrid*

Conventional power grids struggle to accommodate the intermittency of renewable energy sources. In contrast, urban microgrids—operating on a "distributed generation + intelligent dispatch" model—enable a transition in energy supply from "centralized control" to "multi-source coordination." The core of this system involves integrating distributed energy resources (DERs) such as solar photovoltaics (PV) and small-scale wind power with energy storage systems and controllable loads, while using a unified platform to optimize energy flow. This approach not only increases the hosting capacity for renewables but also enhances regional grid stability through peak shaving and valley filling.

Another critical value of microgrids lies in improved energy resilience. When faults occur in the main grid, a microgrid can switch to island mode, operating independently to maintain a continuous power supply for critical facilities like hospitals and data centers. Furthermore, this "local generation for local consumption" model redefines the role of end-users: they evolve from passive consumers to active "prosumers" who participate in energy production via rooftop PV and other means, thereby fostering a decentralized energy ecosystem.

5.5.3 *Dynamic Environmental Monitoring*

A key inefficiency in environmental governance is the lag time and limited accuracy in identifying pollution sources. High-resolution mobile monitoring technology addresses this by enabling a paradigm shift through "dynamic sensing and intelligent analysis." Unlike the static data from fixed monitoring stations, sensor-equipped mobile platforms—such as electric ride-hailing vehicles and inspection drones—can cover broader areas and detect subtle spatial variations in pollution. This dynamic approach compensates for the spatial blind spots of fixed stations, allowing authorities to promptly pinpoint exceedance zones rather than relying on area-wide controls based on averaged data. The core value of mobile monitoring lies in establishing a closed loop of "pollution source tracing, measure implementation, and effectiveness verification." When the system detects an abnormal rise in emission concentrations on a road segment, it can analyze integrated traffic flow and meteorological data to determine if the cause is vehicle exhaust, subsequently triggering immediate measures like temporary traffic restrictions or enhanced street cleaning. Post-intervention data feedback then validates the effectiveness of these actions, creating a cycle of continuous optimization. This model represents a significant advance over traditional methods. For example, mobile monitoring in industrial zones can accurately differentiate the emission contributions of individual factories, providing a basis for differentiated reduction targets and substantially improving the cost-effectiveness of environmental governance.

5.5.4 Urban Carbon Accounting

The scientific rigor of carbon emission reduction hinges on accounting accuracy. The Urban Carbon Accounting Framework (UCAF) overcomes the limitations of traditional statistical methods by integrating multi-source data. Its core objective is to construct a full-life cycle carbon footprint tracking system, encompassing emissions from energy production (e.g., coal-fired power generation, industrial processes), consumption activities (e.g., building operations, transportation), and waste management (e.g., methane emissions). By cross-validating multi-dimensional data—including IoT sensor readings, corporate reports, and remote sensing observations—UCAF enables the creation of a dynamically updated carbon emission map. Unlike traditional approaches reliant on sampling statistics, UCAF can pinpoint the carbon emission profiles of specific city blocks, industrial sectors, and even individual buildings.

5.6 Challenges and Prospects for Intelligent Cities

5.6.1 Data Semantic Heterogeneity

Data, as a core component of intelligent city, faces a critical obstacle to breaking down departmental silos: semantic interoperability. This heterogeneity stems not merely from format differences, but from fundamental discrepancies in how different departments define and measure the same concepts. Compounding the issue is the long-entrenched "data inertia" within departments, which further perpetuates these disparities. Legacy systems often persist with non-standardized data storage formats, with little incentive for modernization. Concurrently, inter-departmental tensions over "data ownership" create resistance to adopting unified semantic standards. This dilemma is not technologically insurmountable—solutions like cross-departmental data dictionaries exist. The true challenge lies in overcoming administrative fragmentation and competing interests. Without addressing these underlying issues, data integration [18] risks remaining superficial, failing to enable meaningful synergy, as depicted in Fig. 5.4.

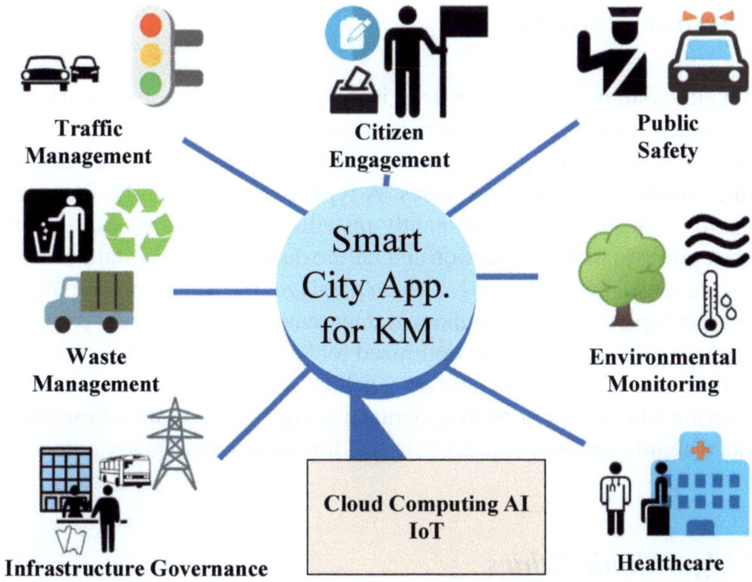

Fig. 5.4 Examples of intelligent city applications and technologies where knowledge management is involved

5.6.2 Privacy Protection

The extensive data dependency of intelligent city creates an inherent tension with the imperative of citizen privacy protection. On the one hand, cross-analyses of multi-source data are prerequisite for the technologies to deliver their intended benefits. Intelligent healthcare, for instance, relies on the integration of electronic health records, physical examination data, and lifestyle information to enable chronic disease risk prediction. On the other hand, richer data dimensions increase the risk of personal re-identification through what might be termed "data linkage" of seemingly disconnected fragments, inevitably sparking privacy concerns. Privacy protection currently faces two major practical dilemmas. The first is data utility loss due to over-anonymization. To mitigate risks, key fields such as age and residential address are often entirely removed from resident data, leaving only meaningless codes. This renders the data useless for policymaking—for example, preventing the identification of mobility patterns specific to elderly populations. The second dilemma involves the ambiguous boundaries of contextual privacy. Traffic cameras monitoring vehicle flows inevitably capture facial images of roadside pedestrians. There are no unified standards governing whether or how such "incidentally collected" private data should be anonymized.

5.6.3 *System Resilience*

Intelligent city systems often demonstrate insufficient resilience, a critical vulnerability that becomes starkly apparent during extreme weather events or cyber incidents. From a technical standpoint, a primary cause is that most systems are designed with inadequate consideration for edge cases. A typical example can be found in smart transportation: sensors are predominantly installed on road surfaces, making them highly susceptible to water immersion and failure during heavy rainfall. This can lead to a city-wide sensor network outage, which paralyzes the intelligent dispatch system and forces a complete reliance on manual intervention. Similarly, the dispatch algorithms of smart grids are primarily optimized for "conventional load profiles." When confronted with extreme temperatures—such as a sharp spike in air-conditioning demand during a heatwave or heating demand during a cold snap—these algorithms can produce significant errors, potentially leading to regional power outages.

5.6.4 *Algorithmic Ethics*

Algorithmic decision-making in intelligent city may embed discriminatory biases that are difficult to detect due to the black-box nature of many algorithms. This ethical risk primarily arises from two sources. First, biases can originate from the training data itself. Since algorithms derive their decision-making logic from historical data, any pre-existing societal biases within that data—such as underrepresentation of women in employment records—can be perpetuated and even amplified. Second, goal-oriented bias occurs when algorithms are optimized for a single, narrow objective at the expense of other values. For example, a traffic dispatch algorithm designed solely to "reduce vehicle delay" might compromise the rights of pedestrians and non-motorized vehicles by narrowing sidewalks or encroaching on bicycle lanes. More critically, the opacity of these systems makes it challenging to audit and trace the root causes of such ethical failures. Ultimately, these multi-dimensional dilemmas stem from a fundamental misalignment between the "technical logic" of intelligent city systems—which prioritizes efficiency and standardization—and the "social logic" of urban life, which is inherently diverse, complex, and dynamic.

5.7 Conclusion Remarks

This chapter elaborates on the Scenarios Engineering for Services Science (SE4SS) framework applied to intelligent city, emphasizing the integration of perception, decision-making, and execution within a sustainable urban ecosystem. The discussion highlights the transition from technology aggregation to eco-digital synergy, where digital twins serve as the operational core enabling dynamic urban governance, environmental sustainability, and resilience.

At the systemic level, intelligent city leverages multimodal sensing and AI-driven modeling to achieve real-time, high-fidelity urban representations. Adhering to FAIR data principles ensures interoperability and transparency across heterogeneous data sources, while privacy-by-design mechanisms safeguard citizen rights. The decision-making architecture integrates Multi-Agent Systems (MAS), Large Language Models (LLMs), and Reinforcement Learning (RL) to form a closed-loop framework of task decomposition, strategy generation, and adaptive adjustment, thereby enhancing urban autonomy and coordination.

The digital twin architecture extends from single-domain applications to city-scale integration, constructing a real-time mapping between physical and virtual systems. Through dual-driven "mechanism + data" modeling, cross-domain interoperability, and virtual–real feedback loops, digital twins evolve into a comprehensive platform for simulation, prediction, and policy validation.

In energy and environmental governance, intelligent systems transition from reactive remediation to proactive regulation. Techniques such as model predictive control for building energy optimization, urban microgrids for distributed energy management, mobile sensing for dynamic pollution monitoring, and multi-source urban carbon accounting collectively enable precise and adaptive environmental management.

Finally, the chapter identifies four persistent challenges: semantic heterogeneity in data, privacy and ethical risks arising from intensive data use, system resilience under extreme or unexpected conditions, and algorithmic bias stemming from opaque decision models. Addressing these issues requires not only technological innovation but also institutional and ethical alignment to ensure that intelligent city evolves as trustworthy, inclusive, and adaptive ecosystems.

References

1. Yeon, H., Eom, T., Jang, K., et al. (2023). DTUMOS, digital twin for large-scale urban mobility operating system. *Scientific reports, 13*(1), 5154.
2. Reichstein, M., Camps-Valls, G., Stevens, B., et al. (2019). Deep learning and process understanding for data-driven Earth system science. *Nature, 566*(7743), 195–204.
3. Jetz, W., Cavender-Bares, J., Pavlick, R., et al. (2016). Monitoring plant functional diversity from space. *Nature Plants, 2*(3), 1–5.
4. Creutzig, F., Roy, J., Lamb, W. F., et al. (2018). Towards demand-side solutions for mitigating climate change. *Nature Climate Change, 8*(4), 260–263.
5. Fromer, J. C., & Coley, C. W. (2024). An algorithmic framework for synthetic cost-aware decision making in molecular design. *Nature Computational Science, 4*(6), 440–450.
6. Rolf, E., Proctor, J., Carleton, T., et al. (2021). A generalizable and accessible approach to machine learning with global satellite imagery. *Nature Communications, 12*(1), 4392.
7. Grantz, K. H., Meredith, H. R., Cummings, D. A. T., et al. (2020). The use of mobile phone data to inform analysis of COVID-19 pandemic epidemiology. *Nature Communications, 12*(1), 4961.
8. Aleta, A., Martin-Corral, D., Pastore, Y., Piontti, A., et al. (2020). Modelling the impact of testing, contact tracing and household quarantine on second waves of COVID-19. *Nature Human Behaviour,4*(9), 964–971.

9. Wu, K., Ding, J., Lin, J., et al. (2025). Big-data empowered traffic signal control could reduce urban carbon emission. *Nature Communications, 16*(1), 2013.
10. Wilkinson, M. D., Dumontier, M., Aalbersberg, I. J. J., et al. (2016). The FAIR guiding principles for scientific data management and stewardship. *Scientific Data, 3*(1), 1–9.
11. Bongiorno, C., Zhou, Y., Kryven, M., et al. (2021). Vector-based pedestrian navigation in cities. *Nature Computational Science, 1*(10), 678–685.
12. Islam, R., Bose, R., Roy, S., et al. (2025). Decentralized trust framework for smart cities: A blockchain-enabled cybersecurity and data integrity model. *Scientific Reports, 15*(1), 23454.
13. Ma, C., Li, A., Du, Y., et al. (2024). Efficient and scalable reinforcement learning for large-scale network control. *Nature Machine Intelligence, 6*(9), 1006–1020.
14. Fei, N., Lu, Z., Gao, Y., et al. (2022). Towards artificial general intelligence via a multimodal foundation model. *Nature Communications, 13*(1), 3094.
15. Li, M., Liu, C., Pan, X., et al. (2025). Digital twin-assisted graph matching multi-task object detection method in complex traffic scenarios. *Scientific Reports, 15*(1), 10847.
16. Bettencourt, L. M. A. (2024). Recent achievements and conceptual challenges for urban digital twins. *Nature Computational Science, 4*(3), 150–153.
17. Tariq, R., Mohammed, A., Alshibani, A., et al. (2024). Complex artificial intelligence models for energy sustainability in educational buildings. *Scientific Reports, 14*(1), 15020.
18. Maciá-Lillo, A., Mora, H., Jimeno-Morenilla, A., et al. (2025). AI edge cloud service provisioning for knowledge management smart applications. *Scientific Reports, 15*(1), 32246.

Chapter 6
SE4SS for Intelligent Logistics

6.1 Introduction

Amid deepening global trade and the exponential growth of e-commerce, the logistics sector has evolved from a backend support function into a core source of competitive advantage and value creation. Traditional logistics models, reliant on manual planning and human decision-making, excel in handling standard scenarios and maintaining flexibility in negotiations and exception handling. However, as market complexity increases and demand grows for timeliness and transparency, these models reveal significant limitations: high operational costs, proneness to error, inefficient resource allocation, slow response to disruptions, and information asymmetry across the supply chain. These challenges are driving the industry toward AI-powered intelligent transformation, which is fundamentally enhancing the efficiency, reliability, and adaptability of supply chain operations.

The rise of intelligent logistics represents the deep integration and application of artificial intelligence, Internet of Things (IoT), big data analytics, and automation technologies within the logistics domain. It manifests as autonomous warehouses, smart transportation management systems, predictive demand planning, and real-time visibility platforms, as shown in Fig. 6.1. Its revolutionary breakthrough lies in achieving unprecedented levels of operational efficiency, agility, and cost reduction. Intelligent systems can process vast amounts of data in real time to optimize delivery routes, predict potential delays, automate warehouse sorting, and dynamically adjust inventory levels, thereby minimizing waste and maximizing resource utilization. The ability to provide end-to-end visibility and predictive insights allows businesses to proactively manage their supply chains, transforming them from cost centers into strategic assets. Despite significant advantages, the development and deployment of intelligent logistics face considerable challenges. The complexity of integrating heterogeneous systems (IoT sensors, ERP, WMS, TMS) creates significant technical hurdles. Ensuring the security and privacy of vast amounts of sensitive supply chain data is paramount. Algorithmic decision-making in dynamic environments must be

X. Li and F.-Y. Wang, *Scenarios Engineering for Services Science*,
SpringerBriefs in Service Science, https://doi.org/10.1007/978-3-032-24152-8_6

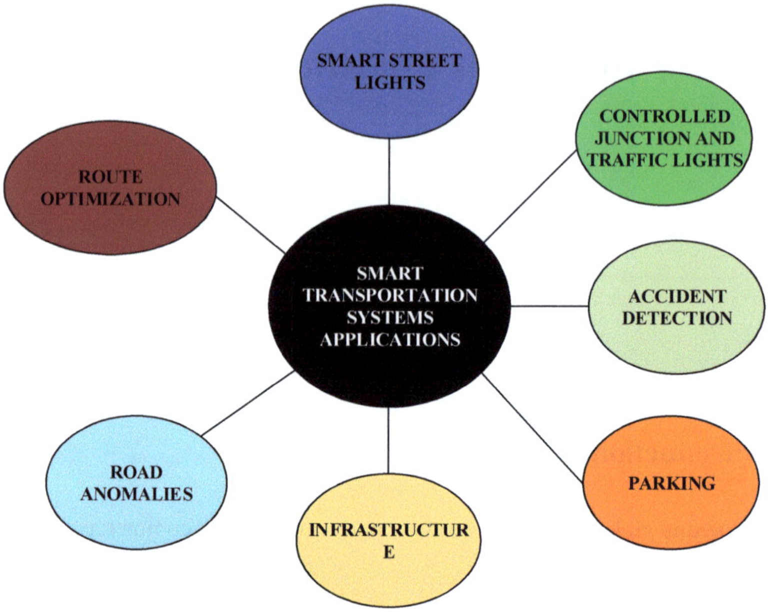

Fig. 6.1 Applications of smart transportation systems

robust, explainable, and fair to avoid unintended consequences. Furthermore, the high initial investment and the need for workforce reskilling present economic and organizational barriers [1].

Scenarios Engineering for Services Science (SE4SS) is a services science methodology that integrates scenarios engineering and new artificial intelligence technology. In the context of intelligent logistics, it is committed to building a highly reliable, resilient, and efficient logistics ecosystem. This methodology systematically constructs multi-level operational scenarios (including actor roles, process flows, environmental variables, and disruption events), combining simulation, predictive analytics, reinforcement learning, and continuous optimization mechanisms. The core of SE4SS is to use scenarios as the "strategic link" for logistics design: on the one hand, through design-oriented scenario visualization, such as digital twins and process simulations, potential bottlenecks and failure points are identified and mitigated; on the other hand, with the help of AI-driven dynamic scenario adaptation, such as real-time rerouting, predictive maintenance, and autonomous resource allocation, the system achieves proactive problem-solving and self-healing capabilities. This chapter mainly introduces the application of SE4SS based on new artificial intelligence for intelligent logistics from the following aspects: IoT-based Intelligent Logistics System; high-quality datasets; foundation models; and advanced model evaluation metrics.

6.2 IoT-Based Intelligent Logistics Systems

Amid the challenges posed by global economic integration and the rapid advancement of computer, communication, and Internet technologies, the scope and speed of logistics activities have entered an unprecedented phase of development. Consequently, intelligent logistics systems characterized by a high degree of informatization and intelligence have emerged. This industry emphasized the comprehensive elevation of informationalization levels, specifically calling for vigorous development of modern logistics, acceleration of a socialized, specialized, and information-based modern logistics service system, and improvement in the management and standardization of logistics. As a result, logistics informatization has become a crucial component of the national information strategy. This study explores the objectives and structure of an IoT-based intelligent logistics system and designs an RFID-based tracking and management system targeted at the manufacturing industry.

6.2.1 System Objectives

Modern Logistics, with the core of informationalization, aims at integrating the logistics resources through sensing, communication, and computer technologies so as to reduce costs and improve service levels. Therefore, the design goal of ILS is to build a modern logistics service system that is automated, visualized, controllable, intelligent, and systematic. It is founded on the core theories and applied technologies of computer networks, communication, acoustics, optics, electronics, mobile computing, GPS, GIS, and artificial intelligence. The system integrates diverse logistics information in real time and performs intelligent data analysis and processing to support efficient operations. With this system, logistics can be seamlessly linked with production. It enables full coordination among business flow, logistics, information flow, and capital flow. These integrations support logistics managers in decision-making and help deliver personalized, value-added logistics services to clients. Intelligent logistics system consists of three layers: interactive sensing layer, network transport layer, application service layer and data service layer, as illustrated in Fig. 6.2.

The interactive sensing layer is responsible for the gathering and convergent processing of material information, in a variety of technologies such as bar code identification, intelligent image processing and identification, RFID (Radio Frequency IDentification), WSN (wireless sensor network), GPS (Global Positioning System), and AIS (Automatic Identification System) which are used for information gathering of different logistics objects. By accessing these interfaces, users can achieve intelligent identification, tracking, positioning, management, and control of objects. The network transport layer serves as the channel for data exchange and transmission, consisting mainly of the Internet, LAN, and wireless communication networks. The Internet functions as the core network of the Internet of Things (IoT), while various

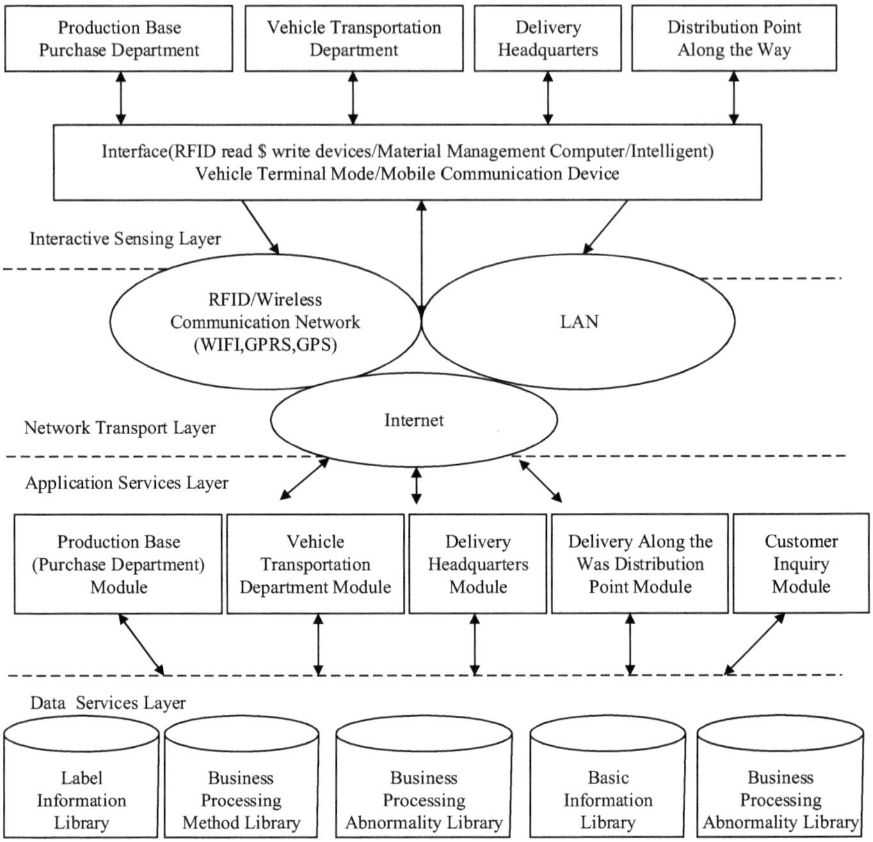

Fig. 6.2 Intelligent logistics system architecture diagram

wireless networks provide anytime, anywhere access. In addition, mobile communication and cluster communication technologies are widely applied in the logistics industry. Application service layer is applied to the services of encapsulating system mainly including electronic data exchange, logistics information system, and so on, with the support of high-performance calculation and mass storage, responsible for the dynamic request at the interactive sensing layer and offering goof information service to clients. Here it is mainly in charge of business service requests of each functional module by different roles. The main technologies used include: cloud calculation, database, mass storage, information retrieval, artificial intelligence and information security protection, etc. Data service layer gathers databases containing different data, offering data sources for the varied applications in the system. The system is supposed to be consisted of a label information database, a basic information database, a business processing method, and a business processing record library.

Users first describe their decision problems. The system then analyzes and identifies the issues according to the administrative tasks and user queries. Next, it builds a user model and performs simplified model operations for evaluation. Through interactive selection and optimization, the most suitable model is chosen. Finally, the system solves and outputs the model results.

6.2.2 *Intelligent Models*

This intelligent model, underpinned by a diverse array of cutting-edge technologies including natural language processing and understanding, advanced image processing, sophisticated data mining, artificial intelligence, and knowledge engineering, leverages a wide range of methodologies. These encompass Association Rules Analysis, evolution analysis, fuzzy logic, genetic algorithm, ant colony algorithm, and neural network algorithm. Additionally, it employs data mining analytic approaches such as classification, clustering, change, and deviation analysis. All these techniques serve to meticulously extract valuable information and underlying principles from the enormous volumes of data generated in logistics operations, thereby providing robust support for business decision-making. Ultimately, this contributes significantly to the realization of substantial value for enterprises, enabling the seamless transformation of raw business data into highly valuable business information and knowledge, as depicted in Fig. 6.3. Such transformation empowers decision-makers to make timely adjustments to market strategies, respond with great agility to clients' ever-evolving needs, effectively reduce operational risks, and substantially enhance the overall levels of decision-making [2].

By establishing various comprehensive logistics decision-making models (encompassing crucial aspects such as logistics information management, logistics transportation optimization, product sourcing strategies, storage management efficiency, decision support mechanisms, reverse logistics processes, etc.), a sophisticated mathematical model is meticulously constructed. The model is designed to minimize delivery frequencies, drastically shorten delivery time, and reduce associated costs to the lowest possible extent. In doing so, it aims to achieve the highly optimized allocation of logistics resources and provide top-notch integrated information system consultation and client services that meet the highest industry standards.

Finally, based on the clearly defined decision-making target, the different granularities of the same index are thoroughly examined from multiple interrelated dimensions of the same perspective. The results derived from these examinations are then effectively integrated using advanced data fusion techniques. This integration process provides a solid and comprehensive foundation of data for multi-dimensional data analysis and data mining in the decision-making process. This comprehensive and rigorous approach ensures that every decision made is not only well informed and data driven but also perfectly aligned with the overarching strategic goals of the logistics system, fostering continuous improvement and innovation within the enterprise's logistics operations [3].

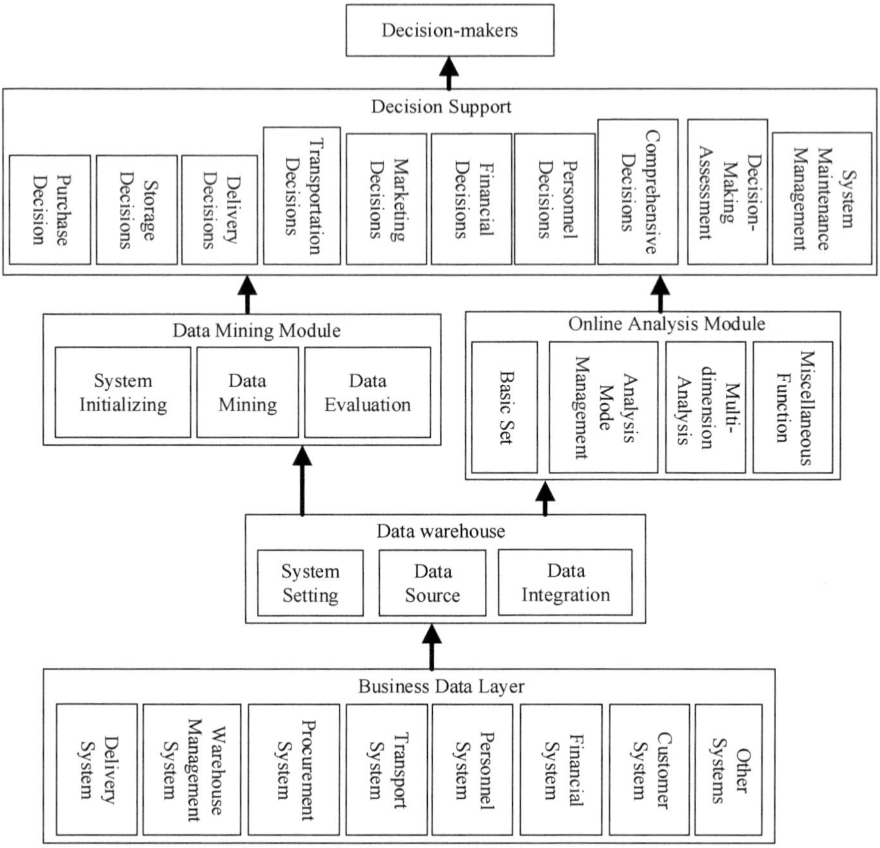

Fig. 6.3 Intelligent logistics system model frame diagram

6.3 New AI-Driven Logistics Service Systems

The IoT-based infrastructure enables large-scale data acquisition and system interconnectivity, forming the foundation of intelligent logistics. However, leveraging artificial intelligence to transform this data into operational services introduces new challenges in ensuring functional safety, prediction stability, and decision-making transparency. In particular, the performance of AI models often diverges significantly between virtual testing environments and real-world deployment, necessitating a deeper understanding of reliability requirements in practical logistics scenarios. This chapter analyzes the reliability challenges faced by AI services in logistics, summarizes current mainstream technological approaches and representative datasets, and compares performance in simulated versus real environments, thereby establishing the necessity of a reliability engineering framework tailored to intelligent logistics.

6.3.1 Reliability Challenges

The development of AI-enabled logistics services relies heavily on virtual-based training and validation to de-risk real-world deployment, yet the inherent divergence between real and virtual environments consistently undermines post-implementation performance. This discrepancy stems not from flaws in AI algorithms themselves, but from the inability of even advanced simulators to replicate the full spectrum of stochastic, systemic, and human-driven variability inherent in logistics operations [4].

Simulations remain irreplaceable for scalable, cost-effective AI testing, as they enable iterative refinement of algorithms without disrupting live supply chains. Leading tools are tailored to address distinct logistics subdomains: AnyLogic, for instance, leverages multi-method modeling (discrete event, agent-based, and system dynamics) to simulate end-to-end e-commerce logistics networks-integrating synthetic data for peak-demand surges (e.g., Black Friday order spikes), last-mile delivery delays, and warehouse labor shift patterns. MATLAB/Simulink, by contrast, excels in hybrid simulations that couple AI route optimization with vehicle dynamics (e.g., electric delivery truck energy consumption under varying load weights), while Siemens Tecnomatix specializes in manufacturing logistics, modeling material flow between production lines and distribution hubs. Reinforcement learning (RL) frameworks further enhance this process: customized OpenAI Gym environments, for example, train warehouse pick-and-place robots to minimize travel time between storage bays, or dynamic inventory RL agents to optimize stock replenishment for perishable goods. Under these controlled conditions, performance metrics are impressive: simulation studies frequently document 15–20% reductions in fleet travel distance, 30% improvements in warehouse loading throughput, and 22% lower inventory holding costs. Critically, however, these outcomes depend on unspoken idealizations: not only perfect communication (e.g., zero-latency data transmission between IoT sensors and AI platforms) and instantaneous data processing, but also static customer demand profiles, standardized package dimensions, and 100% adherence to AI-generated workflows—all of which rarely hold in practice.

Real logistics environments, by contrast, are characterized by cascading complexities that simulations struggle to encapsulate. Sensor inaccuracies, a pervasive issue, manifest in context-dependent flaws: RFID tags on metal pallets suffer 40–60% signal attenuation in high-density warehouses, leading to misread inventory counts; GPS receivers in urban canyons (e.g., downtown Manhattan) experience multi-path errors exceeding 10 meters, disrupting AI-driven last-mile route adjustments. Human interventions introduce further variability: warehouse associates often bypass AI-recommended pick sequences to reduce individual task time, inadvertently increasing overall order fulfillment latency by 15–20%; truck drivers may deviate from AI-planned routes to avoid construction zones unreported in real-time data feeds. Equipment malfunctions, too, transcend simulation simplifications: automated guided vehicles (AGVs) in distribution centers frequently experience unexpected battery drain (due to cold storage temperatures unmodeled in simulations), while conveyor belt jams-triggered by irregularly shaped packages-halt material flow for 30+ minutes, a disruption rarely replicated beyond "binary failure probabilities" in virtual environments, as demonstrated in Fig. 6.4.

More systemic challenges arise from organizational and technological inertia. Legacy enterprise resource planning (ERP) and warehouse management systems (WMS)—still in use by 68% of global logistics firms, per a 2024 Gartner report—often lack APIs for seamless integration with AI platforms, creating temporal lags in data synchronization. For example, a major European parcel carrier found that its 20-year-old TMS delayed vehicle location data transmission to its AI routing system by 90+ minutes, rendering dynamic route adjustments obsolete. Data semantic inconsistencies exacerbate this issue: a warehouse's WMS may define "order completion" as "goods dispatched," while the AI model interprets it as "goods delivered to the customer," leading to misaligned performance metrics and flawed algorithmic learning. External disruptions, meanwhile, introduce unmodeled volatility: unexpected port closures (due to labor strikes), sudden changes in fuel prices, or extreme weather events (e.g., winter storms grounding delivery fleets) are rarely incorporated into simulations with sufficient granularity, as they require real-time, geospatially specific data that simulators cannot precompute [5].

Empirical evidence underscores the severity of this performance gap. A 2024 internal study by DHL analyzed three European parcel distribution centers where AI scheduling algorithms—trained to reduce delivery delays by 28% in simulations—achieved only a 7% reduction in practice, with 32% lower vehicle utilization than projected. Similarly, a port logistics pilot by Maersk found that an AI-driven container loading algorithm, which simulated 25% efficiency gains in virtual tests, delivered just 8% improvement in real operations: tidal fluctuations (underestimated in simulations) delayed vessel docking times, while dockworker shift changes disrupted AI-coordinated loading sequences. Most notably, a cross-industry analysis by the MIT Center for Transportation and Logistics (2023) found that AI logistics tools, on average, experience 25–40% efficiency loss when transitioning from simulation to deployment—with the largest gaps occurring in dynamic environments

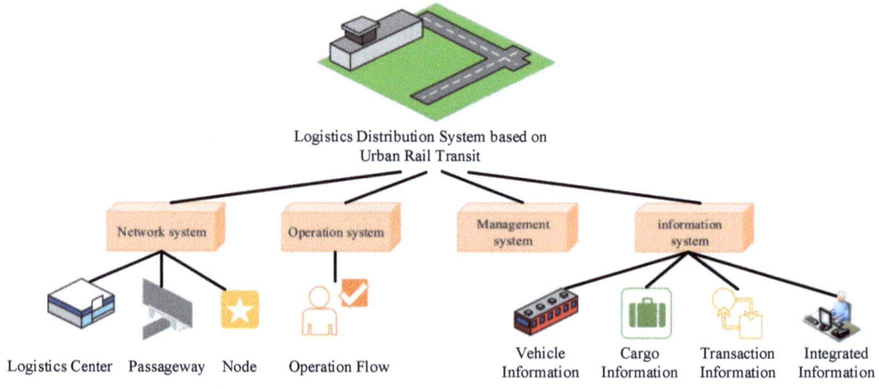

Fig. 6.4 Logistics distribution system based on Urban Rail Transit

(e.g., last-mile delivery) versus more controlled settings (e.g., automated warehouses). These discrepancies reveal a fundamental limitation: simulations excel at modeling predictable complexity, but fail to capture the emergent complexity of human-technology-environment interactions in real logistics systems.

6.3.2 *New AI in Intelligent Logistics*

Time-series prediction models remain fundamental for demand planning and inventory management. Traditional methods such as Seasonal ARIMA and Exponential Smoothing are still prevalent in enterprise software due to their interpretability. However, machine learning approaches like Gradient Boosting (XGBoost, LightGBM) and deep learning architectures (LSTMs, Temporal Convolutional Networks, and Transformers) are increasingly applied for multivariate and high-frequency forecasting. For example, transformer-based models have been used to incorporate external variables (e.g., weather, promotions, social trends) yielding accuracy improvements of 10–15% over traditional models on industrial datasets.

Combinatorial optimization problems inherent in routing, scheduling, and resource allocation are frequently tackled with hybrid approaches. Metaheuristics such as Genetic Algorithms (GA) and Tabu Search are combined with machine learning to guide search processes. Reinforcement learning is gaining traction in dynamic decision-making contexts; for instance, multi-agent RL systems are being developed to coordinate fleet management and warehouse robotics. Companies like Amazon and Alibaba have deployed RL-based systems in fulfillment centers, reporting throughput increases of 12–18%.

Image and video analytics are critical for automated sorting, damage detection, and inventory monitoring. Convolutional neural networks (CNNs)—including architectures like YOLOv5, EfficientDet, and Mask R-CNN—are trained to recognize parcels, pallets, and handling equipment. In practice, vision systems integrated with robotic arms achieve sorting accuracy exceeding 98% in controlled settings, though performance can drop under suboptimal lighting or occlusions [7].

6.3.3 *Datasets and Benchmarks*

The development and validation of AI models in intelligent logistics rely heavily on the availability of realistic and well-structured datasets. These resources serve as critical benchmarks for comparing algorithmic performance and ensuring that proposed solutions can generalize beyond isolated experiments. Publicly available datasets play a fundamental role in academic research and early-stage prototyping. Among the most widely used is the Solomon Dataset, which provides standardized instances for vehicle routing problems (VRP) with varying constraints, customer distributions, and service times, enabling reproducible comparisons of optimization algorithms. For demand forecasting and inventory modeling, the UCI Retail Dataset offers transactional time-series data that supports the development of predictive models under realistic retail conditions. Similarly, the TLC Trip Record Data, drawn from New York City taxi operations, has been adapted for logistics studies to analyze spatiotemporal travel patterns and evaluate route prediction techniques. In the domain of computer vision, datasets such as Pascal VOC and COCO are frequently repurposed for logistics applications, including object detection in warehouse environments and package identification, though they are often supplemented with domain-specific annotations.

In addition to these general-purpose datasets, there is a growing emergence of specialized industrial benchmarks that better capture the complexity and scale of real-world logistics operations. For example, the JD Logistics Prediction Dataset includes large-scale order records, inventory levels, and shipping timelines, supporting advanced research in supply chain forecasting and fulfillment optimization. The AliExpress Logistics Vision Dataset provides annotated images of products and packaging, facilitating the development of robust visual recognition systems for parcel sorting and damage detection. Companies such as DHL and Amazon have also released anonymized operational datasets, which include global freight tracking events and warehouse workflow logs, allowing for more comprehensive testing of AI models in contexts such as delay prediction, exception handling, and robotic coordination.

Despite these valuable resources, significant limitations remain. Most existing datasets are biased toward routine operations and fail to adequately represent edge cases, anomalies, or high-stakes disruption scenarios—precisely the situations where reliability is most critical. Moreover, issues such as label noise, temporal shifts, and sparse annotation often reduce their utility for training trustworthy models. These gaps underscore the need for more sophisticated and representative benchmarks that mirror the full complexity of intelligent logistics systems, including multimodal data integration and cross-system dependencies.

6.4 Key Challenges in Intelligent Logistics

6.4.1 System-Level Issues

The integration of IoT devices such as RFID tags and sensors into logistics assets introduces new operational complexities. A primary concern involves balancing continuous data acquisition and communication with energy conservation. Many IoT devices operate on limited battery power, necessitating energy-efficient protocols to extend operational life without compromising functionality. This calls for advanced energy management strategies and system-level optimizations to ensure that IoT enhancements do not inadvertently reduce the overall sustainability of logistics operations.

Reliable connectivity is essential for IoT systems, especially in logistics environments characterized by mobility and physical obstructions. Metallic structures, thick walls, and large metal containers can significantly attenuate radio signals, leading to packet loss and communication failures. Furthermore, the coexistence of multiple wireless systems within the same frequency band—such as those using ISM bands—can result in interference, further degrading network performance. Designing robust, multi-protocol, and fault-tolerant communication architectures is therefore critical to support seamless data exchange across heterogeneous logistics environments.

The interconnectivity inherent in IoT systems increases their exposure to cyber threats. Data integrity, confidentiality, and access control emerge as major concerns, especially when sensitive logistical information is transmitted across open networks. Emerging technologies such as blockchain show promise in enhancing security through decentralized and tamper-resistant record-keeping. Incorporating such technologies can help mitigate risks related to data manipulation, unauthorized access, and transactional fraud.

Powering IoT devices in remote or mobile settings remains a nontrivial challenge. While batteries are commonly used, they pose safety risks in hazardous environments and entail maintenance costs. Energy harvesting techniques or the use of passive devices may offer alternatives, but a universally applicable power solution has yet to be established. This issue is particularly acute in large-scale deployments involving thousands of devices across global supply chains.

6.4.2 Business Barriers

Deploying an IoT ecosystem requires substantial upfront investment in hardware, software, network infrastructure, and cloud services. Small and medium-sized enterprises (SMEs) may find these costs prohibitive. Additionally, indirect expenses related to system integration, employee training, and process re-engineering contribute to the total cost of ownership. A clear business case demonstrating long-term ROI is essential to justify these investments.

Businesses are often reluctant to share operational data due to concerns over competitiveness and data privacy. This culture of informational isolation inhibits the cross-organizational transparency that IoT systems aim to achieve. Establishing secure, permission-based data-sharing frameworks and clarifying mutual benefits are necessary steps toward fostering trust among supply chain partners.

Integrating new IoT technologies with legacy systems—such as Warehouse Management Systems (WMS) and Transportation Management Systems (TMS)—poses significant technical challenges. Compatibility issues may arise due to differing data protocols, architectural constraints, or operational paradigms, requiring middleware solutions or phased system upgrades to ensure interoperability.

6.4.3 Collaboration Hurdles

Divergent national and regional policies regarding data privacy, communication standards, and freight regulations complicate the implementation of globally integrated IoT-based logistics systems. The absence of international regulatory alignment hinders seamless cross-border operations and data exchange.

A lack of universal standards for IoT devices, data formats, and communication protocols restricts interoperability between systems from different vendors. Widespread adoption of common standards—such as those promoted by international bodies like IEEE or IETF—is essential to achieve scalable and cohesive IoT solutions in logistics.

The adoption of IoT in logistics is not solely a technological issue but a multifaceted challenge involving economic, organizational, and regulatory dimensions. Addressing these barriers requires collaborative efforts among technology developers, logistics service providers, policymakers, and standardization organizations. Future advancements in energy-efficient devices, secure communication architectures, and interoperable platforms will play a pivotal role in overcoming these challenges and unlocking the full potential of IoT in modern logistics systems [8].

6.5 Evaluation Metrics for Intelligent Logistics Models

6.5.1 Basic Metrics

The evaluation of intelligent logistics systems necessitates a comprehensive framework that transcends conventional performance metrics. Unlike standardized service interactions, logistics operations demand quantifiable assessment of precision, resilience, and cognitive adaptability across dynamic supply chain ecosystems.

Precision and Operational Integrity form the cornerstone of logistics AI efficacy. Accuracy manifests in demand forecasting error margins, inventory synchronization fidelity, and route optimization alignment with real-world constraints. Beyond numerical correctness, operational integrity validates end-to-end task completion—ensuring shipments arrive undamaged within contractual timeframes or warehouse robots autonomously reconcile stock discrepancies. Success thresholds must reflect domain-specific criticalities; perishable goods demand stricter on-time delivery tolerances than bulk commodities.

Efficiency bridges algorithmic capability and business impact. Decision latency—the interval between anomaly detection and executable countermeasure generation—must operate at sub-minute resolution for high-velocity supply chains. Process automation rates quantify the proportion of routine decisions resolved without human intervention, directly correlating to operational cost reduction. Concurrently, resource utilization ratios expose optimization gaps between theoretical models and physical execution. Robustness and Fault Tolerance determine system stability amid volatile real-world conditions. Logistics AI must parse unstructured inputs: handwritten shipping labels, contradictory customs documentation, or sensor-faulty IoT data streams. Stress testing simulates adversarial scenarios—sudden disruptions injecting route ambiguity or malware—corrupted inventory records. Quantifiable resilience emerges through consistent performance degradation curves under noise injection and recovery speed metrics after simulated system failures.

The core value of the intelligent Logistics system lies in its ability to accurately and efficiently solve user problems. Accuracy and task success form the primary dimension of evaluating this core competence. Accuracy is typically applied to tasks with clear standard answers, such as intent recognition, entity extraction, and Frequently Asked Questions (FAQ) matching, to measure how well the model output aligns with preset criteria. However, in dynamic, multi-round conversational interactions, the accuracy of literal matching is often not enough to reflect real service performance. The task completion rate goes a step further, focusing on whether the user's core demands are substantively addressed at the end of the conversation, such as whether the order modification was successfully completed and whether specific policy questions were clearly answered. Therefore, defining fine-grained task success criteria based on specific business scenarios is a key prerequisite for accurately evaluating the practicability of the model.

Knowledge coverage is a basic metric to measure the scope of system services and the depth of support, which reflects the proportion of user questions that can be effectively answered by the system knowledge base or feature set. A system with insufficient coverage will frequently encounter out-of-domain (OOD) or have to reject user requests, leading to reduced service availability and increased user frustration. Continuously monitoring OOD or request rejection rates, combined with user feedback to analyze high-frequency uncovered problem areas, is the core basis for driving knowledge base expansion and skill set optimization. High coverage is the basis for ensuring that users can obtain extensive service support through intelligent customer service channels, avoiding frequent manual transfers, and truly giving full play to the scale benefits of automated services.

6.5.2 Advanced Metrics

Network coverage defines the operational scope across multimodal corridors. Coverage gaps manifest when AI cannot interpret region-specific regulations (e.g., biodegradable packaging mandates in the EU) or lacks data for hazardous material routing protocols. Monitoring out-of-scope request frequency drives knowledge graph expansion into edge cases like temperature-controlled pharma logistics or cross-border tariff compliance. Multi-Dimensional Synergy Evaluation elevates assessment beyond transactional metrics. Inter-system fluency measures API-driven harmony between AI and legacy warehouse management (WMS), transportation (TMS), and ERP systems, where natural language alerts must maintain machine-readable precision. Proactive adaptation capability benchmarks predictive interventions—anticipating typhoon disruptions to preemptively reroute fleets or redistributing inventory based on drought—induced crop failures. Stakeholder empathy quantifies clarity in crisis communication, such as generating multi-language delay notifications with compensation options tailored to contractual obligations [9].

Cognitive depth exposes the system's reasoning sophistication. Constraint-based optimization capability resolves competing objectives—minimizing carbon footprint while maintaining 24-h delivery promises amid fuel shortages. Causal reasoning dissects failure root causes: identifying customs clearance delays as downstream effects of blockchain ledger desynchronization. Cross-process coherence ensures shipment tracking data integrity cascades through inventory ledgers and financial reconciliation modules, triggering automated variance resolution protocols. Real-time knowledge grounding via Retrieval-Augmented Generation (RAG) architectures anchors decisions to live regulatory databases, preventing hazardous route suggestions during wildfire embargoes.

This multi-layered evaluation paradigm transforms logistics AI from reactive tools into anticipatory partners, balancing algorithmic rigor with supply chain resilience across increasingly chaotic global networks.

6.6 Conclusion Remarks

This chapter systematically examines the application of SE4SS in intelligent logistics, addressing the industry's demand for enhanced efficiency, reliability, and adaptability amid global trade deepening and e-commerce growth. It contextualizes the shift from traditional manual logistics—plagued by high costs, errors, inefficient resource allocation, and supply chain information asymmetry—to AI-powered systems. By integrating IoT, big data, automation, and AI, intelligent logistics turns supply chains into strategic assets. The chapter details the IoT-based Intelligent Logistics System, including its goal of an automated/intelligent framework and AI-driven models that convert raw data into decision support.

A core focus is AI logistics reliability: while simulations enable cost-effective training (e.g., 15–20% reduced fleet travel distance), simulated-physical discrepancies (sensor inaccuracies, human intervention, legacy lags, unmodeled disruptions) cause 25–40% post-deployment efficiency loss. It also summarizes mainstream AI methods (time-series prediction, reinforcement learning, computer vision) and datasets, noting gaps in edge-case/disruption coverage. Key adoption challenges are outlined: (1) systemic issues (IoT energy, signal attenuation, cybersecurity); (2) business barriers (SME costs, data-sharing reluctance, legacy integration); (3) collaboration hurdles (divergent regulations, no universal IoT standards). A comprehensive evaluation framework is proposed, with basic metrics (precision, efficiency, robustness) and advanced metrics (network coverage, cognitive depth).

Fundamentally, this chapter establishes an SE4SS framework for intelligent logistics, emphasizing reliable AI deployment needs tech innovation (improved simulations, edge-case datasets), collaboration, and standards. Future efforts should prioritize edge-case datasets, AI-legacy integration, and global regulatory alignment to unlock SE4SS's potential for resilient logistics ecosystems.

References

1. Ivanov, D., Dolgui, A., & Sokolov, B. (2019). The impact of digital technology and industry 4.0 on the ripple effect and supply chain risk analytics. *International Journal of Production Research, 57*(3), 829–846.
2. Mohammad, S., Jagendra, S., & Abbas, R. H. (2021). A novel algorithm for capacitated vehicle routing problem for smart cities. *Symmetry, 10*, 1923–1933.
3. Rasmus, R., Christoph, S., & Frederik, F. (2022). Method for systematic assessment of mobile network coverage for logistic applications on the German highway. *Network, 2*, 311–328.
4. Zhu, M., & Huang, Y. P. (2019). The development of intelligent manufacturing drives the rise of intelligent logistics. *Forum on Science and Technology in China, 24*, 62–65.
5. Wang, M. (2023). Design and supply chain management of intelligent logistics system using cloud computing under internet of things. *International Journal of Grid and Utility Computing, 14*, 216–228.
6. Garcez, S. D. A., & Lamb, L. (2020). Neurosymbolic AI: The 3rd wave. *Artificial Intelligence Review, 56*, 12387–12406.
7. Danchuk, V., Comi, A., & Weiß, C. (2023). The optimization of cargo delivery processes with dynamic route updates in smart logistics. *Eastern European Journal of Enterprise Technologies, 3*, 64–73.
8. Chowdhery, A., Narang, S., Devlin, J., et al. (2023). Palm: Scaling language modeling with pathways. *Journal of Machine Learning Research, 24*(240), 1–13.
9. Sijin, L., Yu, L., Zhe, L., et al. (2022). Vision-Guided object recognition and 6D pose estimation system based on deep neural network for unmanned aerial vehicles towards intelligent logistics. *Applied Sciences, 13*, 115–125.

Chapter 7
SE4SS for Intelligent Healthcare

7.1 Introduction

Healthcare is a major concern for the general public in contemporary society. However, the traditional healthcare model is facing multiple challenges. In the diagnosis process, over-reliance on individual doctors' experience has led to a series of issues, including diagnostic errors, treatment delays, and inadequate communication between patients and doctors. Because healthcare resources are unevenly distributed, patients in remote areas often struggle to access medical services and lack proper medical equipment. In addition, the healthcare system faces challenges such as inefficient inspection processes, low accuracy in diagnosing complex diseases, and slow processing of large-scale medical data.

To address the numerous issues of the traditional healthcare model and meet people's demand for reliable healthcare service system, SE4SS (Scenarios Engineering for Services Science) has emerged. Intelligent healthcare encompasses multiple critical aspects, including disease diagnosis, treatment, health management, and drug research and development, offering immense potential for enhancing the quality and efficiency of healthcare service system. SE4SS integrates scenarios engineering and New AL technologies, employing the ACP (Artificial societies, Computational experiments, and Parallel execution) to construct a systematic and comprehensive methodology and framework. SE4SS addresses various dimensions, such as modeling and analysis of complex scenarios, construction and management of datasets, development and evaluation of foundation models, design and coordination of AI agents, realization and assurance of fully autonomous services, and advancement and application of knowledge automation. This ensures that AI can deliver reliable, efficient, and secure services in intelligent healthcare scenarios. This chapter will discuss the application of SE4SS in intelligent healthcare scenarios, providing references and insights for promoting reliable service system in intelligent healthcare.

X. Li and F.-Y. Wang, *Scenarios Engineering for Services Science*,
SpringerBriefs in Service Science, https://doi.org/10.1007/978-3-032-24152-8_7

7.2 Healthcare Scenarios Modeling

7.2.1 Real Healthcare Scenarios Modeling

Real-world scenarios refer to the complex environments that intelligent healthcare directly faces in practical applications, which contain numerous unforeseen factors and challenges, imposing extremely high demands on the reliability, accuracy, and adaptability of intelligent healthcare systems.

From a spatial perspective, medical institutions of different scales and levels vary significantly. Large Grade III Class A hospitals are equipped with advanced facilities and a gathering of experts, capable of handling various difficult and complicated diseases; primary community hospitals, on the other hand, focus on the diagnosis and treatment of common diseases and basic health management. Geographically, medical resources are concentrated in urban hospitals, while relatively scarce in rural areas. From a temporal perspective, the progression of diseases changes rapidly. Acute illnesses such as myocardial infarction and stroke have sudden onset and critical conditions, requiring immediate treatment against time. Chronic diseases like diabetes and hypertension have a long course and need long-term monitoring and management. Meanwhile, the medical process involves multiple links, from patient registration, consultation, examination, diagnosis to treatment and rehabilitation, and each link may be delayed or deviated due to various factors. Individual differences among patients are also an important factor. Age, gender, genes, living habits, etc., all affect disease manifestations and treatment responses. For example, the physical functions of the elderly decline, resulting in poor drug tolerance; patients with different genotypes may have drastically different efficacy and side effects in response to the same drug.

There are various methods for modeling real-world scenarios. The common data-driven modeling integrates multi-source data from HIS (Hospital Information System), EMR (Electronic Medical Record), PACS (Picture Archiving and Communication System), etc., applies data mining techniques to discover potential patterns and rules in the data, and constructs disease prediction, diagnosis, and treatment recommendation models through machine learning algorithms such as decision trees and support vector machines. In addition, tools like flowcharts and Petri nets can be used to sort out and model medical processes in detail, identifying key nodes, bottlenecks, and potential risk points in the processes. Through simulation, the effects of different process improvement schemes are evaluated, such as optimizing the registration process and shortening the waiting time for examinations, so as to improve the efficiency and quality of medical service system. Besides, multi-agent modeling regards doctors, nurses, patients, medical equipment, etc., in the medical system as agents, each with its own goals, behavior rules, and decision-making capabilities. By defining interaction rules and communication mechanisms among agents, complex dynamic behaviors in the medical system are simulated. For instance, simulating the communication between doctors and patients, the process of nurses executing medical orders, etc., to analyze the collaboration efficiency of medical teams and

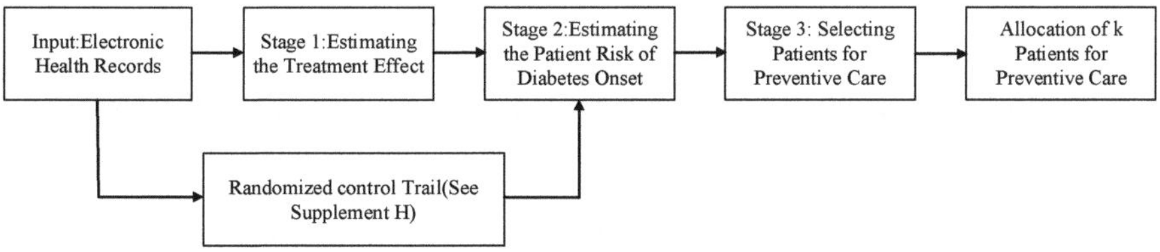

Fig. 7.1 Decision model for data-driven allocations of preventive care

patients' medical experience. Fig. 7.1 shows a data-driven decision model proposed by researchers for allocating preventive care. It is a decision model that receives EHR (Electronic Health Records) from patients as input, with the output being a cost-effective allocation of preventive care resources, as shown in Fig. 7.1. The decision model combines causal reasoning, machine learning, and optimization, and is divided into three consecutive stages for assessing patients' risk of diabetes [1].

7.2.2 Synthetic Healthcare Scenarios Modeling

In intelligent healthcare, high-quality medical data is crucial for training and validating artificial intelligence models. However, real-world medical data is often plagued by issues such as privacy protection and acquisition difficulties, especially the scarcity of data on rare diseases and extreme cases. The generation of medical scenarios can make up for this deficiency, providing richer and more diverse data samples for model training and testing. Additionally, synthetic scenarios can simulate medical situations under various hypothetical conditions, assisting researchers in exploring new treatment methods and strategies, thereby enhancing the model's generalization ability and capability to handle complex situations.

There are various methods for synthetic scenario modeling. The traditional physics-based modeling constructs physiological and pathological models of the human body based on medical knowledge and physical principles. For example, a fluid dynamics model of the cardiovascular system can be established to simulate blood flow in blood vessels, facilitating the study of the mechanisms underlying the occurrence and development of cardiovascular diseases. By adjusting model parameters, simulated data under different disease conditions can be generated to provide references for disease prediction, diagnosis, and treatment. Statistical modeling is another common approach. It analyzes the statistical characteristics of medical data, such as distribution patterns and correlations, and uses statistical methods to generate virtual data that conforms to specific distributions. For instance, based on factors like patients' age, gender, and medical history, the incidence and symptom manifestations of a certain disease can be statistically analyzed, and then virtual cases with similar characteristics can be generated to assist in medical research and training. Furthermore, deep learning-based modeling has gained prominence. It leverages deep learning models such as GANs (Generative Adversarial Networks) and VAEs (Variational Autoencoders) to learn the distribution characteristics of real medical data (e.g., electronic medical records, medical images) and generate realistic synthetic data. These synthetic data can be used to augment training datasets, especially for rare diseases where real data is scarce, to improve the performance of diagnostic and treatment models. Virtual medical scenarios are simulated medical environments created using technologies such as Virtual Reality (VR) and Augmented Reality (AR). In virtual medical scenarios, medical staff can conduct surgical simulation training and disease demonstration, while patients can perform rehabilitation training and psychological therapy. Artificial Intelligence (AI) can enable real-time monitoring and analysis of the virtual environment in these scenarios, providing personalized guidance and recommendations to healthcare professionals and patients. However, there are certain differences between virtual medical scenarios and real-world medical scenarios, and AI services need to consider how to transfer the experience and skills acquired in virtual scenarios to real-world settings. Additionally, the performance of virtual medical devices and user experience can also impact the effectiveness of AI services, necessitating continuous optimization and improvement. A medical knowledge graph represents and stores knowledge in the medical field in a graphical format, clearly illustrating the relationships between medical concepts. Through technologies such as Natural Language Processing (NLP) and Machine Learning (ML), medical knowledge graphs can be automatically generated from vast amounts of data, including medical literature and electronic health records. The generated medical knowledge graphs can provide rich knowledge support for intelligent medical services, such as disease diagnosis and treatment recommendation. However, the process of generating a medical knowledge graph is complex, requiring the handling of large volumes of heterogeneous data while ensuring the accuracy and completeness of the knowledge. Moreover, as medical knowledge is constantly evolving, AI services need to promptly update and maintain medical knowledge graphs to reflect the latest medical advancements.

7.3 Generative Medical AI Models

7.3.1 Transformer Model

Medical foundation models represent a significant application of large-scale pretrained model technology in the healthcare domain and have become a key research focus in intelligent medical assistance. By leveraging pretraining on vast amounts of medical data, these models exhibit critical capabilities such as cross-task transfer, multimodal understanding, and complex reasoning, overcoming several limitations of traditional neural networks in medical applications. With these capabilities, medical foundation models are reshaping the implementation of core tasks such as assisted diagnosis, clinical report generation, and medical image analysis. They hold profound implications for achieving general intelligence in healthcare. BERT (Bidirectional Encoder Representations from Transformers), through its bidirectional encoding mechanism, can deeply understand the contextual semantic information in medical texts. It is pre-trained on large-scale unannotated medical corpora, acquiring rich linguistic knowledge and domain-specific medical features, and can then be fine-tuned for specific medical tasks such as medical question answering and medical record classification. The GPT (Generative Pre-trained Transformer) series models are renowned for their powerful text generation capabilities; GPT-3 and subsequent versions can generate high-quality, coherent medical-related texts, which can assist in writing medical reports, creating health science articles, and more.

7.3.2 *Stable Diffusion Model*

In terms of image processing, image generation models like Stable Diffusion show great potential. They can generate realistic medical images based on text descriptions, helping doctors understand patients' conditions more intuitively and being applicable to medical education and simulated surgical scenarios. Additionally, there are deep learning foundation models specifically designed for medical imaging, such as those for chest X-ray and CT image diagnosis. These models can automatically detect lesion features, identify disease types, and provide diagnostic references for doctors. In medical text analysis tasks, BERT models are widely used for the classification and information extraction of electronic medical records. For instance, they can classify electronic medical records by disease type to facilitate quick retrieval and statistical analysis; extract key information from medical records, such as patients' symptoms, examination results, and diagnostic conclusions, to build structured medical databases. A research team utilized BERT models to analyze a large number of electronic medical records of lung cancer patients, accurately extracting information on patients' pathological features and treatment plans, which provided strong support for clinical research and personalized treatment. In medical image diagnosis, foundation models based on deep learning have achieved remarkable results. Taking ophthalmic disease diagnosis as an example, some foundation models can automatically analyze fundus images to detect early signs of common ophthalmic diseases such as diabetic retinopathy and glaucoma. By comparing with the diagnostic results of professional ophthalmologists, these models have achieved a high level of accuracy, significantly improving the screening efficiency of ophthalmic diseases. Fig. 7.2 shows a comparison chart of deep learning-based cardiac MRI (Cardiac Magnetic Resonance Imaging) reconstruction effects, demonstrating the application of foundation models in the field of medical imaging [2].

In the field of disease prediction, foundation models, combined with multi-source data such as patients' historical medical data, genetic information, and lifestyle, can predict the risk of patients developing certain diseases in the future. For example, foundation models are used to predict high-risk groups for cardiovascular diseases, enabling early intervention measures to reduce the incidence of the disease. Fig. 7.3.

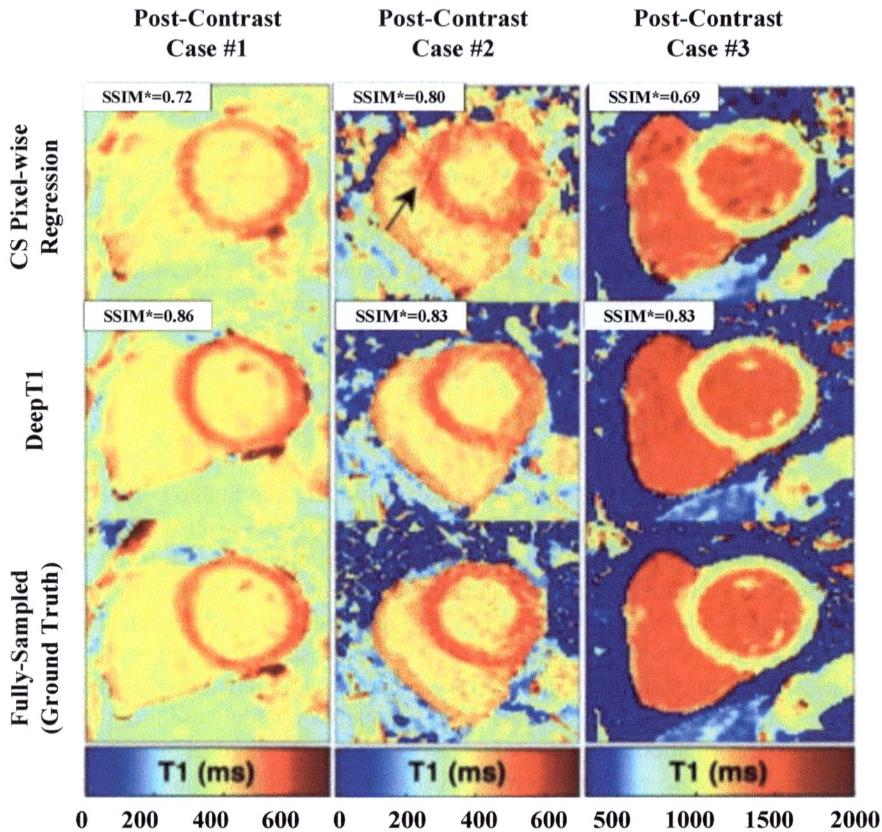

Fig. 7.2 Comparison chart of deep learning-based cardiac MRI (Cardiac Magnetic Resonance Imaging) reconstruction effects

illustrates the workflow for clinical decision support using LLM (Large Language Model). This workflow shows using LLMs in clinical decision support for referral, triage, and diagnosis. The workflow begins with the input general user input and clinical user input—followed by prompts engineered to predict triage, specialty, and diagnosis. The predictions are generated using the Claude family models and a RAG-assisted Claude 3.5 Sonnet model. These predictions [3] are then evaluated and compared with the ground truth to assess the performance of the LLMs.

Common metrics for evaluating the performance and reliability of artificial intelligence models cover multiple aspects. Accuracy is one of the most fundamental metrics, representing the proportion of samples correctly predicted by the model, and is suitable for scenarios with balanced classes (e.g., the binary classification of pneumonia/non-pneumonia). However, in healthcare, the proportion of positive samples for rare diseases (e.g., amyotrophic lateral sclerosis) may be less than 1%, where high accuracy might mask the model's ability to identify minority classes.

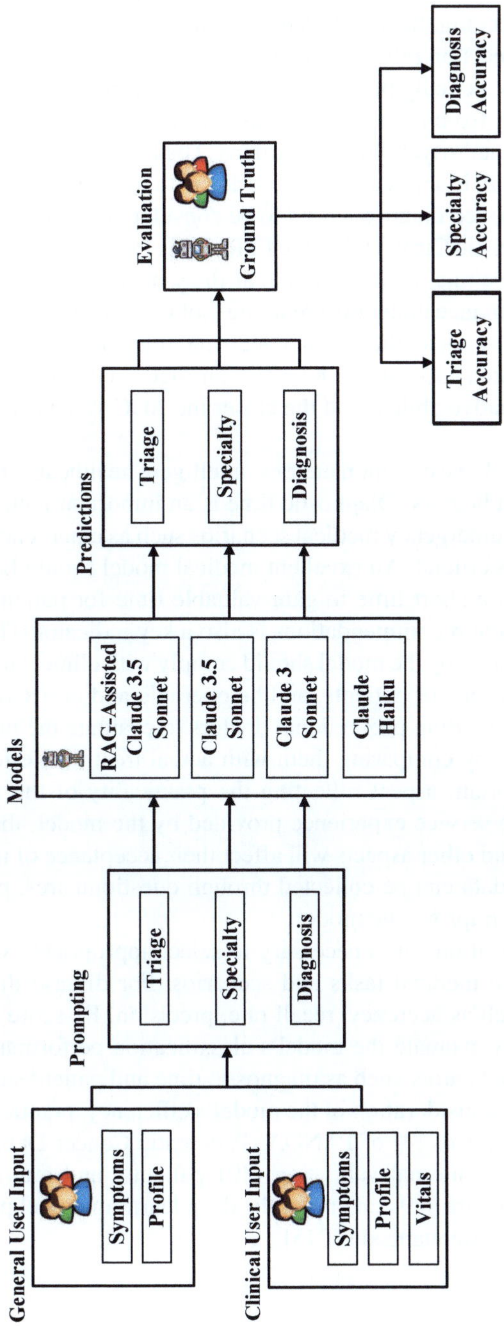

Fig. 7.3 Workflow for clinical decision support using LLM (Large Language Model)

Recall rate indicates the proportion of actually positive samples that are correctly predicted, reflecting the model's capability to avoid missed diagnoses. For example, in breast cancer screening, the recall rate needs to reach over 95% to reduce the risk of late detection. Precision refers to the proportion of samples predicted as positive by the model that are actually positive. In medical scenarios, high precision can prevent healthy patients from being misdiagnosed as ill, thereby reducing unnecessary treatments and patients' psychological burdens. The F1-score, as the harmonic mean of precision and recall, is applicable to scenarios with imbalanced classes where both misdiagnosis and missed diagnosis need to be considered.The ROC curve (Receiver Operating Characteristic Curve) takes the FPR (False Positive Rate) as the x-axis and the TPR (true positive rate, i.e., recall rate) as the y-axis, demonstrating the model's classification performance under different thresholds. In medical scenarios, the ROC curve can help doctors select the optimal decision threshold. The AUC (Area Under Curve) value is the area under the ROC curve; an AUC > 0.9 indicates that the model has strong discriminative ability, and the closer the AUC value is to 1, the better the model's performance.

In addition to the above common metrics, intelligent healthcare also requires some special evaluation indicators. Diagnostic time is an important metric for measuring model efficiency. In emergency medical scenarios such as emergency care, rapid and accurate diagnosis is crucial. An excellent medical model should be able to provide diagnostic results in a short time to gain valuable time for patient treatment. The rationality of treatment recommendations is also a key indicator. The treatment recommendations provided by the model should comply with clinical medical standards and the actual conditions of patients, and be operable and effective. This indicator can be measured by inviting professional doctors to evaluate the model's treatment recommendations or by comparing them with actual treatment effects. Patient satisfaction is an important aspect reflecting the practicality of the model. Patients' satisfaction with the service experience provided by the model, the explanation of diagnostic results, and other aspects will affect their acceptance of medical services. Patient satisfaction data can be collected through questionnaires, patient feedback, etc., to evaluate and improve the model.

In practical applications, it is necessary to select appropriate evaluation metrics according to specific medical tasks and scenarios. For disease diagnosis models, common metrics such as accuracy, recall rate, precision, F1-score, and AUC value can comprehensively evaluate the model's classification performance. Meanwhile, combining special indicators such as diagnostic time and patient satisfaction allows for a comprehensive consideration of the model's efficiency, practicality, and patient experience. The indicators [4] of PANDA (Pancreatic Cancer Detection with Artificial Intelligence) on internal validation (291 patients) and multi-center external validation (5,337 patients in 9 centers) of the deep learning model pancreatic cancer detection based on non-enhanced CT [5].

7.4 Multimodal Medical Datasets

In the field of intelligent healthcare, high-quality data serves as the cornerstone of SE4SS (Scenarios Engineering for Services Science). The quality, scale, and diversity of such data directly influence the performance and reliability of artificial intelligence models. Take datasets for radiopathology and proteogenomics as an example. Liu et al. constructed a high-quality dataset containing multimodal data. The study collected MRI (Magnetic Resonance Imaging), pathological, whole-exome sequencing, RNA sequencing, and proteome data from 1,194 patients, and also collected GBM (Glioblastoma Multiforme) data from public databases for validation, providing comprehensive data support for multimodal fusion analysis [6].

7.4.1 Dataset Sources

Core medical systems are important sources of medical data, including HIS (Hospital Information System), EMR (Electronic Medical Record), PACS (Picture Archiving and Communication System), LIS (Laboratory Information System), and other multi-source systems. These systems integrate diverse data such as patients' basic information, medical records, medical images, and examination data. By leveraging data mining techniques, potential patterns and rules within the data can be identified. Machine learning algorithms like decision trees and support vector machines are then used to build models for disease prediction, diagnosis, and treatment recommendations. Additionally, tools such as flowcharts and Petri nets can be employed to map and model medical processes in detail, identifying key nodes, bottlenecks, and potential risks. Through simulation, the effectiveness of different process improvement schemes can be evaluated—for instance, optimizing registration procedures or reducing examination waiting times—to enhance the efficiency and quality of medical services. Moreover, multi-agent modeling treats doctors, nurses, patients, and medical equipment in the healthcare system as agents, each with its own goals, behavioral rules, and decision-making capabilities. By defining interaction rules and communication mechanisms between agents, complex dynamic behaviors in the healthcare system can be simulated. For example, simulating the communication between doctors and patients or nurses executing medical orders to analyze the collaboration efficiency of medical teams and patients' treatment experiences. Public health data provides macro-level information for medical research, including disease prevalence trends, population health status, and vaccination coverage. Research data encompasses a wealth of medical research findings and clinical trial data, which are instrumental in driving innovation and advancement in medical technology. Data collected by wearable devices, such as heart rate, blood pressure, sleep quality, and step count, facilitate real-time monitoring of patients' health status, particularly for individuals with chronic diseases in their daily health management.

In terms of data types, medical data can be categorized into structured, semi-structured, and unstructured data. Structured data, such as patients' age, gender, and examination indicators, has a fixed format and clear semantics, playing a vital role in statistical analysis of medical data and the construction of disease prediction, diagnosis, and treatment recommendation models. Semi-structured data, like medical records in XML format, has a certain structure but is not fully standardized, requiring parsing and processing to extract useful information. Unstructured data, such as medical images and text-based medical records, lacks a fixed format and structure, making processing more challenging. Medical images require image processing techniques for feature extraction and analysis, while text-based medical records demand natural language processing to understand their semantic information.

7.4.2 Dataset Life Cycle

Data collection is the primary step in building high-quality datasets. To obtain comprehensive and accurate data, it is necessary to integrate multiple data sources, including data from different hospitals, departments, and medical records over various time periods. During the collection process, strict adherence to data privacy protection regulations is imperative to ensure patients' personal information is not leaked. The collected data must undergo data cleaning, which involves removing redundant data (e.g., duplicate patient information and repeated test results), handling erroneous data (e.g., abnormal age values and incorrect units for test indicators), and filling in missing data (using methods like mean or median imputation). Data annotation is a critical method in improving data quality. For structured data, unified standards for classification and coding are required to ensure consistency and comparability. For unstructured data, such as medical images and text reports, professional medical personnel are needed for annotation. For example, in medical image annotation, doctors must accurately label the location, size, and morphology of lesions to provide accurate labels for training artificial intelligence models. To enhance annotation efficiency and accuracy, methods such as multi-person annotation, cross-validation, and quality review of annotation results can be adopted. Storing medical data requires consideration of its massive volume, rapid growth, and high security requirements. Distributed storage systems can effectively address the scalability of data storage by dispersing data across multiple nodes, improving data read-write performance and reliability. Meanwhile, to ensure data security, multiple encryption technologies should be employed for encrypted storage, preventing data theft or tampering during transmission and storage. For instance, a combination of symmetric and asymmetric encryption can be used for data encryption and decryption.

High-quality datasets form the foundation for training reliable artificial intelligence models. By learning from large volumes of real medical data, AI models can capture various disease characteristics and patterns, thereby improving the accuracy of disease diagnosis and the reliability of treatment outcome predictions. For example, in medical image diagnosis, training AI models using large datasets of annotated images enables them to recognize different types of lesions (e.g., tumors and inflammations) and provide accurate diagnostic results.

7.5 AI Agents for Intelligent Healthcare

An AI agent is an intelligent entity that perceives its environment and autonomously takes actions based on perceived information to achieve specific goals. In intelligent healthcare, it exhibits a high degree of autonomy, operating according to preset rules and algorithms without the need for continuous human intervention. It also possesses adaptability, allowing it to adjust behavioral strategies as the environment and data change, and interactivity, enabling effective communication with patients, medical staff, and other components of the healthcare system. Fig. 7.4. demonstrates the use of AI agents to empower biomedical research. AI agents pave the way for "AI scientists" capable of skeptical learning and reasoning. These multi-agent systems consist of agents based on conversable LLMs (Large Language Models) and can coordinate ML (Machine Learning) tools, experimental platforms, humans, or even combinations of them [7].

7.5.1 Types of Intelligent Healthcare AI Agents

From a functional perspective, AI agents can be categorized into reactive, deliberative, and hybrid types. Reactive AI agents respond directly to environmental stimuli without engaging in complex reasoning or planning, characterized by rapid responsiveness. Examples include simple medical alarm devices that immediately sound an alert when detecting a patient's vital signs exceeding normal ranges. Deliberative AI agents, on the other hand, have internal states and reasoning capabilities, enabling them to plan and make decisions based on environmental information and their own goals. For instance, an intelligent diagnostic AI agent synthesizes information such as a patient's symptoms, medical history, and test results, and uses medical knowledge and reasoning algorithms to derive diagnostic conclusions. Hybrid AI agents combine the features of reactive and deliberative types, adopting different processing methods in various situations to enhance their ability to handle complex environments. In terms of application scenarios, there are medical consultation AI agents, surgical assistance AI agents, rehabilitation care AI agents, etc. Medical consultation AI agents can answer patients' common health questions and provide medical advice and popular science knowledge. Surgical assistance AI agents can offer services such as image navigation and surgical instrument positioning for doctors during operations. Rehabilitation care AI agents can guide patients in rehabilitation training and monitor training effects.

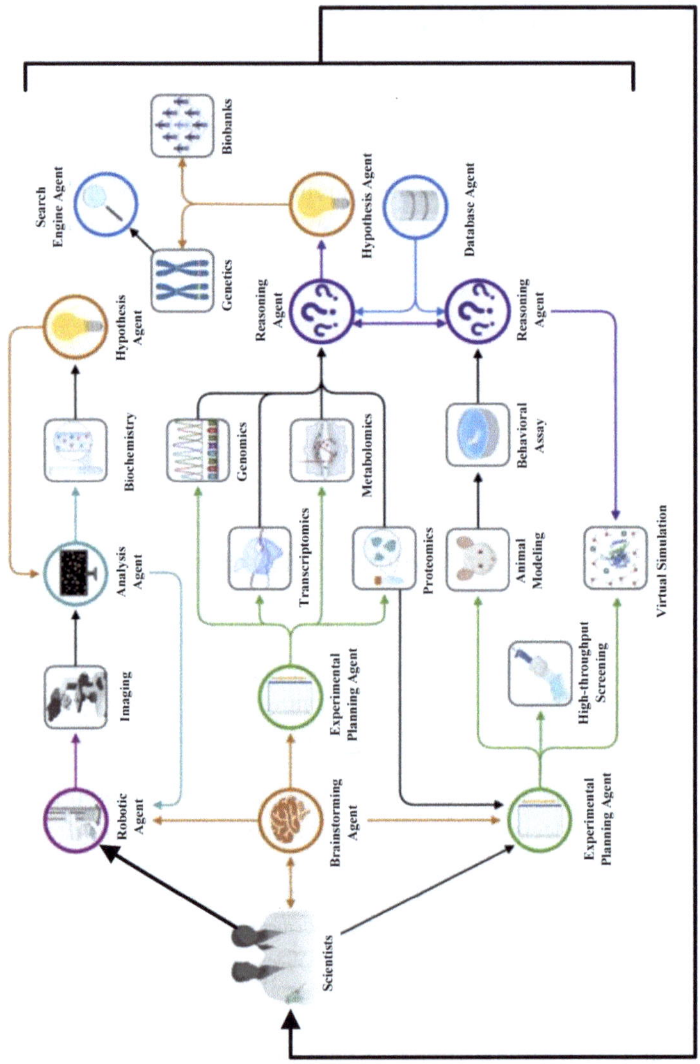

Fig. 7.4 Use of AI agents to empower biomedical research

7.5.2 Challenges of Intelligent Healthcare AI Agents

AI agents face numerous challenges in intelligent healthcare applications. Data quality and privacy issues are critical: medical data comes from diverse sources, has various formats, and contains a large amount of sensitive information, making it a major challenge to ensure data accuracy, completeness, and security. The interpretability of models is also poor; doctors and patients often struggle to understand the basis for decisions made by AI agents, which limits their clinical application to a certain extent. Additionally, the legal and ethical responsibilities of AI agents are not clearly defined, making it difficult to assign liability in the event of medical accidents. In the future, AI agents will develop toward greater intelligence, personalization, and humanization. With the advancement of multimodal data fusion technology, AI agents will be able to comprehensively utilize multiple types of data such as voice, images, and text, improving their ability to understand and handle complex medical situations. Research on explainable AI technology will be further advanced to make the decision-making process of AI agents more transparent and understandable. Meanwhile, AI agents will deeply integrate with technologies such as the Internet of Things and blockchain, building a safer and more efficient intelligent healthcare ecosystem to provide patients with higher-quality and more convenient medical service system.

7.6 Conclusion Remarks

This chapter delves into the SE4SS framework for intelligent healthcare, aiming to address various challenges faced by traditional medical models, such as diagnostic errors, treatment delays, uneven resource distribution, and low efficiency in data processing. The SE4SS integrates scenarios engineering with new artificial intelligence technologies and adopts the ACP method (Artificial societies, Computational experiments, Parallel execution) to construct a systematic and comprehensive methodology, ensuring the efficiency, reliability, and safety of intelligent healthcare service system.

In terms of complex scenario modeling, this chapter elaborates on the modeling methods for real-world scenarios and synthetic scenarios. Real-world scenario modeling enhances the personalization and precision of medical service system by integrating multi-source data, applying data mining and machine learning technologies to build models for disease prediction, diagnosis, and treatment recommendations. Meanwhile, synthetic scenario modeling technologies, especially deep learning-based GAN (Generative Adversarial Network) and VAE (Variational Autoencoder) models, provide rich and diverse data samples for model training, strengthening the model's generalization ability and capability to handle complex situations. Regarding foundation models and evaluation metrics, this chapter introduces various large models that play a key role in the field of intelligent healthcare, such as BERT (Bidirectional Encoder Representations from Transformers), GPT series, and Stable

Diffusion. It also discusses in detail multiple metrics for evaluating model performance and reliability, including accuracy, recall, precision, F1-score, ROC curve, and AUC (Area Under Curve) value. These metrics provide a scientific basis for model optimization and clinical application. As the cornerstone of the SE4SS, the construction and management of high-quality datasets are crucial. This chapter emphasizes the key steps of data collection, cleaning, annotation, and storage, pointing out that distributed storage systems and multiple encryption technologies should be adopted to ensure data security, integrity, and reliability. As core components of intelligent healthcare, AI agents have brought revolutionary changes to medical services with their autonomy, adaptability, and interactivity. This chapter categorizes and discusses reactive, deliberative, and hybrid AI agents, and looks forward to their application prospects in medical consultation, surgical assistance, rehabilitation care, and other fields. At the same time, it also points out the challenges faced by AI agents in data quality, model interpretability, and legal ethics.

In summary, the SE4SS provides comprehensive and in-depth methodological support for reliable services in intelligent healthcare. With the continuous advancement of technology and the continuous expansion of application scenarios, the SE4SS framework will play an increasingly important role in improving the quality, efficiency, and patient experience of medical services. In the future, with the in-depth integration of multimodal data fusion, explainable artificial intelligence, and technologies such as the Internet of Things and blockchain, the intelligent healthcare ecosystem will become safer and more efficient, bringing higher-quality and more convenient medical services to patients.

References

1. Kraus, M., Feuerriegel, S., Saar-Tsechansky, M. (2023). Data-driven allocation of preventive care with application to diabetes mellitus type II. *Manufacturing & Service Operations Management, 26*, 1.
2. Oscanoa, J. A., Middione, M. J., Alkan, C., et al. (2023). Deep learning-based reconstruction for cardiac MRI: A review. *Bioengineering, 10*(3).
3. Gaber, G., Shaik, M., Allega, F., et al. (2025). Evaluating large language model workflows in clinical decision support for triage and referral and diagnosis. *Digital Medicine, 8*.
4. Liu, X., Liu, H., Yang, G., et al. (2025). A generalist medical language model for disease diagnosis assistance. *Nature Medicine, 31*, 932–942.
5. Cao, K., Xia, Y., Han, X., et al. (2023). Large-scale pancreatic cancer detection via non-contrast CT and deep learning. *Nature Medicine, 29*, 3033–3043.
6. Liu, Z., Wu, Y., Xu, H., et al. (2025). Multimodal fusion of radio-pathology and proteogenomics identify integrated glioma subtypes with prognostic and therapeutic opportunities. *Nature Communications, 16*.
7. Gao, S., Fang, A., Huang, Y., et al. (2024). Empowering biomedical discovery with AI agents. *Cell*.

Chapter 8
SE4SS for Intelligent Vehicles

8.1 Introduction

Intelligent vehicles are an interdisciplinary technological framework. They use advanced artificial intelligence and systems engineering to provide highly reliable, secure, and sustainably optimized service systems for intelligent vehicles operating in complex environments. Developed in response to the inadequacies of conventional automotive architectures—particularly their inability to satisfy the stringent demands of Level 3+ autonomy in real-time processing, fault tolerance, and adaptive functionality. This paradigm leverages advanced AI techniques such as deep learning, reinforcement learning, and federated learning, while contending with persistent challenges including model opacity, data privacy concerns, and generalization in rare edge cases.

At its core, an intelligent vehicle leverages reinforcement learning for adaptive decision-making through vehicleâŁ“simulation co-training [1]. It applies deep learning with multimodal fusion and spatiotemporal modeling to achieve high perception and prediction accuracy. Large-scale digital twin simulations are used to create comprehensive edge-case datasets. The entire process is supported by formal verification and AIOps monitoring, ensuring a robust trainâŁ“validateâŁ“deploy life cycle. Practical implementations include cloudâŁ“vehicle collaborative intelligence, personalized driving strategies, and predictive maintenance systems. However, the field still faces major challenges, such as vulnerability to adversarial attacks, latency in edge computing, unclear ethical standards [2], and complex regulatory diversity [3]. Future progress is expected to emphasize neuro-symbolic systems that combine logical reasoning with deep learning, foundation models for generalized autonomy, and quantum-accelerated simulation and verification. Together, these advances aim to build a fully autonomous, self-optimizing perceptionâŁ“decisionâŁ“service pipeline, ensuring operational excellence and trustworthiness under all driving conditions.

X. Li and F.-Y. Wang, *Scenarios Engineering for Services Science*,
SpringerBriefs in Service Science, https://doi.org/10.1007/978-3-032-24152-8_8

Future research will likely focus on neuro-symbolic systems that combine logical reasoning with deep learning, foundation models for generalized autonomy, and quantum-accelerated simulation and verification. Together, these innovations aim to realize a fully autonomous, self-optimizing perceptionâŁ“decisionâŁ“service pipeline that delivers operational excellence and trustworthiness under all driving conditions.

8.2 Complex Traffic Scenarios Modeling

The development and implementation of intelligent vehicle technology rely heavily on large-scale, data-driven approaches. At the core is the synergistic integration of real-world data collection and virtual simulation platforms.

8.2.1 Real Traffic Scenarios

In real environments, multi-sensor systems—including LiDAR, cameras, and other perception modules—continuously capture road topology, traffic behaviors, and environmental conditions under varying light and weather, as shown in Fig. 8.1. These datasets provide essential ground truth for training perception algorithms and optimizing decision-making models. However, physical vehicle testing alone cannot meet the speed and scale required for algorithm iteration due to high costs, regulatory constraints, and the difficulty of reproducing rare or long-tail scenarios.

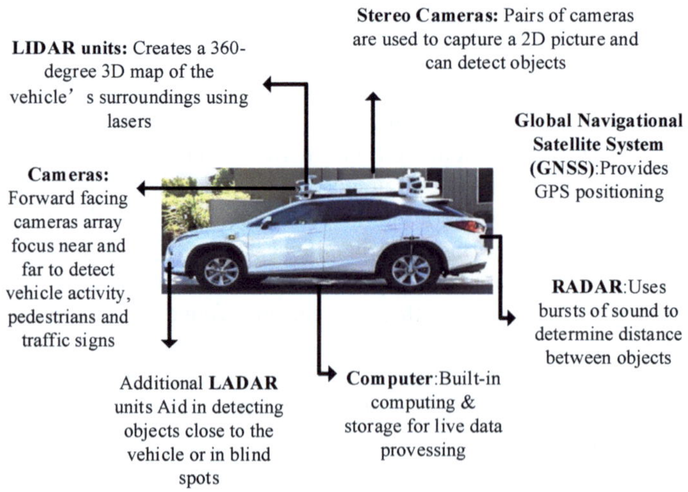

Fig. 8.1 High-level functional parts of a standard intelligent autonomous vehicle system

8.2.2 Synthetic Traffic Scenarios

By constructing parameterized digital twin environments, researchers can precisely control variables such as weather, traffic flow, road geometry, and unexpected events. This approach enables the creation of systematic test libraries that include both normal and extreme conditions. Synthetic data-based validation allows accelerated testing equivalent to tens of thousands of kilometers while safely exploring hazardous edge cases that are rarely encountered in real driving [5]. Virtual simulation platforms further enhance realism and transferability by aligning physics engines and sensor models with real sensor data. As shown in Fig. 8.2, AI-generated simulation scenarios enable systematic evaluation of vehicleâŁ“behavior interactions under controlled yet diverse traffic conditions. From a systems engineering perspective, combining real and synthetic data creates a complementary dual-track framework. Real-world data ensures generalization, while synthetic data challenges robustness through controlled boundary conditions. This integrated strategy reduces testing risks and costs and forms a closed-loop evaluation cycle, supporting the continuous evolution of perception, prediction, and decision-making modules through iterative feedback. With advances in neural rendering and physics-based modeling, this virtualâŁ“real fusion paradigm is becoming an industry standard, providing vital technical support for the safe deployment of next-generation autonomous vehicles.

8.3 Representative Datasets for Intelligent Vehicles

During the training and validation of intelligent vehicles systems for intelligent vehicles, diverse datasets corresponding to various road conditions and driving scenarios are employed to enhance model robustness and generalization. These datasets capture a wide spectrum of real-world driving environments, traffic behaviors, and edge cases to ensure comprehensive learning and evaluation. Below, we introduce several commonly used datasets in the intelligent vehicles domain.

8.3.1 ApolloScape Dataset

The ApolloScape dataset is an open-source dataset designed for intelligent vehicles research, encompassing various aspects such as trajectory prediction, 3D LiDAR object detection and tracking, scene parsing, lane segmentation, self-localization, 3D vehicle instances, stereo vision, and inpainting. It aims to foster innovation and advancement in intelligent vehicles technology.

Fig. 8.2 Diagram of AI-synthesized driving scenario simulation

8.3.2 NuScenes Dataset

The nuScenes dataset [6] represents a large-scale intelligent vehicles dataset comprising 3D bounding box annotations for 1,000 scenes collected in Boston and Singapore. Each scene spans a duration of 20 s with annotations provided 2 Hz frequency. The dataset contains a total of 28,130 samples for training, 6,019 samples for validation, and 6,008 samples for testing purposes. It features a comprehensive sensor suite for autonomous vehicle applications, including a 32-beam LiDAR system, six cameras, and radars with full 360-degree coverage, thereby providing multimodal perception capabilities for robust environmental understanding.

8.3.3 OpenLane-V2 Dataset

The world's first intelligent vehicles benchmark for road structure perception and reasoning. The primary objective of this dataset is to facilitate scene structure perception and reasoning, which requires models to accurately identify the drivable status of lanes in the surrounding environment. The tasks encompassed in this dataset extend beyond lane centerline and traffic element detection to include the recognition of topological relationships among detected objects, thereby enabling a comprehensive understanding of complex driving scenarios.

8.3.4 4Seasons Dataset

The 4Seasons dataset is a large-scale benchmark for visual SLAM and long-term localization in intelligent vehicles. Featuring over 300km of recordings across nine environments, it provides centimeter-accurate global poses for evaluating visual odometry, place recognition, and visual localization, with special focus on seasonal and weather variations to advance intelligent vehicles research.

8.3.5 NuPlan Dataset

The nuPlan dataset serves as a closed-loop reactive simulation benchmark for evaluating end-to-end intelligent vehicles models. It segments long-distance driving journeys into multiple manageable scenarios, each containing high-level navigation information and corresponding sensor data. This dataset primarily addresses the limitations of existing simulators in terms of authenticity and reactivity, particularly in highly interactive driving situations.

8.4 Advanced Modules for Intelligent Vehicles

Engineers have extensively employed AI-assisted methodologies to participate in the design of various intelligent vehicle modules, which has significantly enhanced the efficiency of intelligent vehicles development. Specifically, AI technologies have been integrated into multiple critical phases, including requirements analysis, algorithm development, data engineering, model optimization, and verification testing.

8.4.1 Dataset Processing

During the data acquisition phase, a reinforcement learning-based sampling algorithm analyzes scene complexity in real time. It dynamically adjusts LiDAR scanning frequency and camera resolution, improving data collection efficiency by over 40%. In the data annotation stage, a multimodal fusion semi-supervised learning framework achieves 95% of the accuracy of full supervision while using only 10% of manually labeled samples. An active learning algorithm further enhances efficiency by automatically detecting hard cases and tripling the focus of annotation resources. For data processing, a generative adversarial network-powered augmentation system synthesizes photorealistic extreme weather scenarios, expanding long-tail scene datasets by 5–8 times. The quality validation process employs an anomaly detection neural network that automatically identifies labeling errors and sensor anomalies, demonstrating 60% higher detection rates compared to conventional manual inspection methods.

8.4.2 Visual Perception and Understanding

During the requirement definition phase, a spatiotemporal graph convolutional network analyzes large volumes of driving video data. It identifies key visual perception requirements through behavioral pattern analysis, greatly accelerating the demand assessment process. In the algorithm architecture design phase, a neural architecture search system with multi-objective optimization automatically generates hybrid-scale attention networks. These architectures balance computational efficiency, power consumption, and detection accuracy, showing significant improvements over traditional design methods. For instance, Bao et al. [7] introduced Heat-Assisted Detection and Ranging (HADAR), which addresses the challenge of textureless ghost images in thermal vision recognition while establishing performance benchmarks through AI-enhanced thermal sensing. This HADAR technology not only achieves texture and depth perception through darkness with daylight-like clarity but also extends its sensing capability to discern ambiguous physical properties beyond conventional RGB or thermal vision, thereby paving the way for fully passive and physics-aware machine perception systems. Specifically, the paper proposes a TeX image processing methodology. It degrades and decomposes recognized images to improve analysis. This approach systematically processes visual data through mathematical transformations, enabling more robust feature extraction and pattern analysis in complex imaging scenarios. The technique demonstrates particular effectiveness in handling noise interference and multi-layer information separation, offering improved computational efficiency compared to conventional image processing pipelines while maintaining reconstruction accuracy.

8.4.3 Path Planning and Vehicle Control

AI significantly enhances system safety and adaptability by establishing a comprehensive perception-decision-execution intelligent assistance framework. At the environmental modeling level, a spatiotemporal graph neural network predicts interactions among traffic participants in real time. It dynamically deciphers relationships between vehicles, pedestrians, and other entities, accurately capturing their intention evolution patterns. This enables the planning module to achieve proactive scenario understanding and make more informed decisions. To improve real-time responsiveness in dynamic environments, neural differential equation technology is applied. It enables smooth trajectory optimization across continuous state spaces, producing kinematically feasible motion paths under strict time constraints.

8.4.4 Testing Metrics for Intelligent Vehicles

Evaluating traditional intelligent vehicles models is a multifaceted process. It mainly assesses accuracy, safety, reliability, and efficiency across key functions, including environmental perception, behavior prediction, path planning, and motion control.

At the perception level, key metrics evaluate the model's accuracy in recognizing vehicles, pedestrians, traffic signs, and lane markings. They include mean Average Precision (mAP) for object detection, mean Intersection over Union (mIoU) for semantic segmentation, and lateral offset error for lane detection. For behavior prediction, metrics center on the Average Displacement Error and Final Displacement Error to gauge the accuracy of forecasting the future trajectories of surrounding objects. For vehicle planning and control, key indicators focus on trajectory tracking performance. They include lateral position error and heading angle deviation, as well as comfort metrics such as acceleration and jerk. Safety is assessed through collision rate, time-to-collision, and minimum distance to obstacles. Ultimately, at the system level, overall performance is evaluated through extensive simulation or real-road testing, with core metrics including disengagement rate, system-level collision rate, and Operational Design Domain compliance rate. Collectively, these metrics form the foundational framework for assessing whether an intelligent vehicles system is sufficiently safe, reliable, comfortable, and compliant with traffic regulations. However, traditional metrics exhibit limitations in comprehensively quantifying complex interactive behaviors and the capability to handle extreme scenarios.

8.5 New Artificial Intelligence for Intelligent Vehicles

The intelligent vehicles field is undergoing rapid advancement yet faces several critical challenges. A major challenge is ensuring safety and reliable decision-making in complex, dynamic real-world environments [8]. This is especially difficult in rare situations, such as extreme weather, unexpected obstacles, and highly interactive traffic participants. Core technical hurdles include overcoming sensor limitations under adverse conditions, reducing excessive reliance on high-definition maps, and enhancing algorithmic capabilities for prediction and planning in intricate traffic situations. High hardware costs still limit large-scale adoption. The lack of global regulatory standards and clear liability frameworks for accidents also creates major challenges. In addition, public trust in fully driverless technology remains low, posing important policy and societal barriers.

Looking ahead, development priorities focus on leveraging large-scale AI models and deep multi-sensor fusion to elevate environmental comprehension and intelligent decision-making, exploring cooperative vehicle-infrastructure-cloud intelligence to extend perception boundaries and optimize traffic efficiency, advancing more cost-effective technological pathways like pure vision solutions, and fostering cross-industry collaboration to accelerate the maturation and standardization of regulations and insurance models. The ultimate objective is to achieve commercial deployment in logistics and shared mobility scenarios, progressively reshaping the future transportation ecosystem while rigorously upholding safety and ethical principles. Current research frontiers include Neural Radiance Fields (NeRF) and World Models as key areas of investigation.

NeRF: NeRF represents a novel deep learning-based approach for reconstructing and rendering complex 3D scenes from sparse 2D image sets. This technique fundamentally models a continuous three-dimensional environment through a large-scale neural network. The network learns to predict volumetric density and color attributes at any spatial coordinate. NeRF simulates the physical process of light rays traversing from a camera through pixels into the scene. Along each ray path, the method samples multiple spatial points, querying density and color attributes at these locations. Volume rendering techniques subsequently integrate these sampled values along the ray, accumulating and synthesizing the final pixel color. By generating view-consistent novel imagery solely from input multi-angle photographs, NeRF achieves high-fidelity 3D scene reconstruction and free-viewpoint novel viewpoint synthesis.

8.5.1 Transformer Architecture

The Transformer represents a revolutionary deep learning architecture for sequence modeling, fundamentally centered on its self-attention mechanism. This mechanism dynamically computes the relevance weights between every token within an input sequence and all other tokens, enabling parallelized capture of long-range semantic dependencies and thereby overcoming the long-term memory limitations inherent in traditional recurrent neural networks. The architecture has three main technical components. First, the self-attention layer uses multi-head attention to capture different semantic relationships, such as syntax and coreference. Second, positional encoding methods, like sinusoidal or cosine functions, add sequence order information to token embeddings to replace the missing recurrent structure. Third, feed-forward neural networks with residual connections apply non-linear transformations to the attention outputs and help stabilize deep network training [11].

Within the large language model (LLM) domain, the Transformer has become foundational. Generative models like Generative Pre-Trained Transformer (GPT) and Large Language Model Meta AI (LLaMA) utilize a decoder-only architecture, leveraging autoregressive generation for text creation and code synthesis. Comprehension-focused models such as Bidirectional Encoder Representations from Transformers (BERT) adopt an encoder-only structure, employing bidirectional attention for semantic understanding and classification tasks. Next-generation hybrid architectures, exemplified by Jamba and GPT-OSS, integrate Transformer layers with state space models and mixture-of-experts systems. These hybrids preserve the global attention advantage while expanding context windows to exceeding 128,000 tokens through sparse computation, significantly enhancing long-context processing efficiency, as demonstrated in Fig. 8.3.

The Transformer's revolutionary impact manifests in three primary dimensions: its parallel computation architecture unlocks the full potential of GPU and TPU hardware, enabling the training of models with hundreds of billions of parameters; the global attention mechanism transcends the distance constraints of prior sequence modeling techniques; its modular design inherently supports multimodal extensions, as seen in Vision Transformers for image processing and models like Whisper for speech recognition. Today, the Transformer not only powers leading foundation models including GPT-4 and Claude but also continues to evolve through innovations like rotary position encoding and sliding window attention, solidifying its role as the core engine driving AI's progression from language mastery toward broader artificial general intelligence.

8.5.2 Vision-Language-Action Models

The integration of foundation models, including LLM, vision-language models (VLM), and vision-language-action models (VLA), with intelligent vehicles systems overcomes the limitations of traditional rule-based algorithms dependent on manually engineered rules and high-definition maps by endowing vehicles with semantic comprehension, logical reasoning, and human-like decision-making capabilities. The core operational workflow involves several critical stages. First, vehicles utilize VLM or VLA to fuse multimodal sensory data from cameras, radar, and other sensors, achieving not only object detection but also deep semantic understanding of environmental contexts and participant intentions. Subsequently, these systems construct dynamic environmental representations that interpret traffic regulations, road conditions, and surrounding vehicle and pedestrian behaviors, preparing for subsequent path planning. Then, leveraging structured knowledge bases, they perform logical reasoning and risk assessment to anticipate potential hazards and execute human-like driving decisions. Finally, VLA translate these decisions into safe operational control commands.

Concurrently, LLM enable natural language interaction, allowing drivers to issue verbal commands or request information directly. Beyond real-time operation, these foundation models automatically generate simulation scenarios to train systems for extreme edge cases and continuously refine performance through real-world data feedback loops. This comprehensive workflow facilitates a paradigm shift in intelligent vehicles from mechanical perception-to-control reactions toward cognitive decision-making intelligence, substantially enhancing vehicular adaptability, operational safety, and interaction naturalness in complex environments.

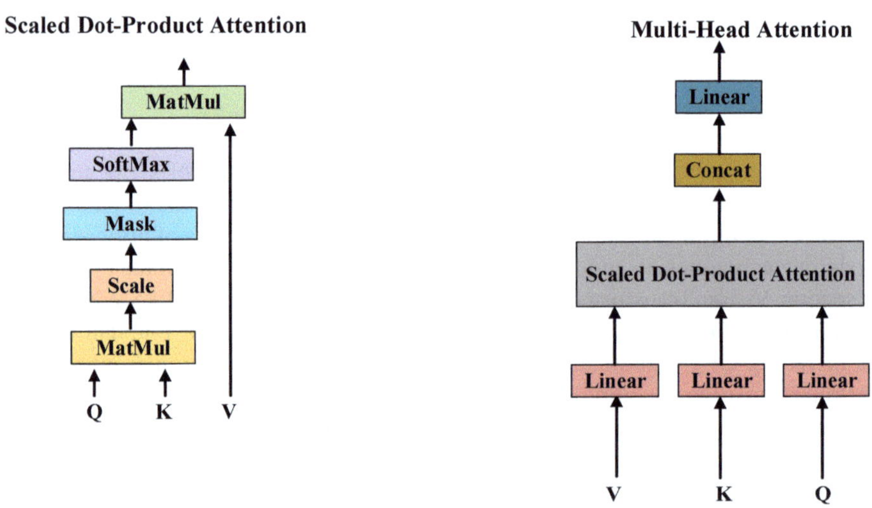

Fig. 8.3 (left) Scaled Dot-Product Attention. (right) Multi-Head attention consists of several attention layers running in parallel

8.5.3 World Models

World Models represent a fundamental concept in artificial intelligence, enabling agents to internally construct and comprehend dynamic representations of their environments. By learning the governing principles of how environmental states evolve over time in response to their own actions, these models can predict probable future state sequences based on current observations and past memory. This predictive capability empowers agents to perform effective reasoning under partial observability, safely conduct trial-and-error learning within their internal models to devise optimal strategies, and better adapt to novel or complex situations. Consequently, world models significantly enhance learning efficiency and decision-making intelligence, establishing themselves as a pivotal pathway toward more general and human-level artificial intelligence systems.

8.5.4 Foundation Models

Foundation model-driven intelligent vehicles is emerging as a transformative technological paradigm within the field, fundamentally centered on leveraging massive datasets and powerful general-purpose artificial intelligence models, such as the Transformer architecture, to unify critical functions including perception, prediction, and decision-making [9, 10]. Distinct from traditional modular approaches reliant on extensive hand-crafted rules and disparate models, foundation models employ end-to-end or tightly coupled methodologies to learn complex environmental comprehension, intention reasoning, and driving policies directly from multimodal sensor inputs encompassing camera images, LiDAR point clouds, and textual commands. This approach demonstrates significantly enhanced adaptability and generalization capabilities for dynamic open-world scenarios and statistically rare edge cases. These models not only achieve more accurate recognition and prediction of surrounding environments and traffic participant behaviors but also enable human-like complex decision-making and planning by integrating commonsense reasoning and experiential knowledge, substantially improving system safety, operational fluency, and human-like driving behavior under unknown or ambiguous road conditions.

Concurrently, foundation models accelerate efficiency in simulation scenario generation, automated data annotation, and continuous system iteration. While challenges persist in real-time performance, model interpretability, and rigorous safety verification, foundation models are unequivocally propelling intelligent vehicles toward greater intelligence, reliability, and the ultimate goal of full autonomy, progressively reshaping the future landscape of transportation.

8.6 Autonomous Intelligence in Intelligent Vehicles

Autonomous intelligence represents emerging conceptual frameworks that are transforming technological development. Within the SE4SS paradigm, autonomous intelligence enables next-generation vehicular systems to achieve self-learning capabilities, automated knowledge processing, and enhanced operational reliability. Through integration with blockchain and smart contracts, these systems can maintain immutable behavioral records, secure vehicleâŁ"infrastructure interactions, and execute consensus-driven operational rules. The synergy between autonomous intelligence and distributed ledger infrastructures facilitates intelligent decision-making, predictive maintenance, and adaptive service optimization in next-generation transportation ecosystems, ensuring transparent, auditable, and verifiable AI-driven behaviors.

8.6.1 Cognitive Architectures and Trust Mechanisms

Autonomous intelligence constitutes fundamental technological pillars in intelligent vehicles, integrating environmental perception, real-time decision-making, and cognitive understanding. By leveraging multi-sensor fusion, advanced AI models, and knowledge graphs, vehicles gain the ability to interpret complex scenarios such as sudden obstacles and temporary traffic regulations. The technical architecture transforms abstract knowledge—including traffic rules, driving experience, and crowdsourced operational insights—into structured semantic networks, enabling logical reasoning and risk anticipation.

Blockchain infrastructures augment this cognitive architecture by offering distributed trust anchors. Smart contracts encode policy constraints, driving protocols, and safety mechanisms into executable logic, preventing unauthorized system manipulation and strengthening compliance in V2V (vehicle-to-vehicle) and V2I (vehicle-to-infrastructure) communication. In this context, decentralized governance models like DAO (Decentralized Autonomous Organization) allow heterogeneous agents—vehicles, infrastructure nodes, service operators—to collectively update safety rules, data-sharing policies, and model parameters under transparent consensus. These mechanisms establish a cooperative intelligent transportation trust layer beyond centralized platforms.

8.6.2 DAO-Driven Operations

Beyond enabling precise control in routine driving, autonomous intelligence offers safety-critical fallback decision-making under extreme weather, sensor malfunction, or adversarial scenarios by accessing its knowledge base. Through automated simulation and closed-loop data optimization, the system continuously improves its ability to handle edge cases, evolving toward full-scenario, highly reliable, and explainable cognitive-driven architectures.

Blockchain-anchored vehicle logs provide tamper-proof histories of perception events, decision trajectories, and system failures, enabling regulatory-compliant auditing and liability attribution. Smart contracts support decentralized service coordination, automatically enforcing maintenance actions, battery swaps, and insurance claims when predefined conditions are met. DAO frameworks foster community-level intelligence: fleets of vehicles collaboratively optimize routing, update scenario models, and negotiate energy allocation in mobility ecosystems. This decentralized intelligence redefines safety boundaries and efficiency paradigms in intelligent mobility, establishing benchmarks for trustworthy autonomous operation. Integrating autonomous intelligence with programmable trust infrastructures ultimately bridges the gap between machine perception and human-level situational understanding, marking a significant leap toward truly intelligent, self-governing transportation ecosystems.

8.7 Conclusion Remarks

This chapter studies the development and challenges of reliable service systems using new AI for intelligent vehicles. It introduces the complexity of driving scenarios in both real and simulated traffic. It also describes simulation datasets for intelligent vehicles training, AI modules for intelligent vehicles service design, and evaluation metrics for industry models. Finally, it explains how AI models connect with autonomous vehicle systems and introduces related ideas such as autonomous intelligence.

As AI keeps improving, intelligent vehicles cares more about safety, reliability, and transparency. Engineers must make sure trustworthy AI is applied in real systems and verified for safety. This means building adaptive service architectures that use foundation models and embodied intelligence to improve vehicle understanding and decision-making. Online learning should keep improving reliability in complex and changing traffic. A unified evaluation framework will help advance reliable service systems with new AI and move closer to full vehicle autonomy.

References

1. Feng, S., Sun, H., Yan, X., et al. (2023). Dense reinforcement learning for safety validation of autonomous vehicles. *Nature, 615*(7953), 620–627.
2. Kuznietsov, A., Gyevnar, B., Wang, C., et al. (2024). Explainable AI for safe and trustworthy autonomous driving: A systematic review. *IEEE Transactions on Intelligent Transportation Systems, 25*(12), 19342–19364.

3. Li, B., Wang, Y., Mao, J., et al. (2024). Driving everywhere with large language model policy adaptation. In *Proceedings of the IEEE/CVF Conference on Computer Vision and Pattern Recognition* (pp. 14948–14957).
4. Tyagi, A. K., & Aswathy, S. U. (2021). Autonomous Intelligent Vehicles (AIV): Research statements, open issues, challenges and road for future. *International Journal of Intelligent Networks, 2*, 83–102.
5. Li, W., Pan, C. W., Zhang, R., et al. (2019). AADS: Augmented autonomous driving simulation using data-driven algorithms. *Science Robotics, 4*(28), eaaw0863.
6. Caesar, H., Bankiti, V., Lang, A. H., et al. (2020). Nuscenes: A multimodal dataset for autonomous driving. In *Proceedings of the Conference on Computer Vision and Pattern Recognition* (pp. 11621–11631).
7. Bao, F., Wang, X., Sureshbabu, S. H., et al. (2023). Heat-assisted detection and ranging. *Nature, 619*(7971), 743–748.
8. Liu, H. X., & Feng, S. (2024). Curse of rarity for autonomous vehicles. *Nature Communications, 15*(1), 4808.
9. Chen, L., Sinavski, O., Hünermann, J., et al. (2024). Driving with llms: Fusing object-level vector modality for explainable autonomous driving. In *2024 IEEE International Conference on Robotics and Automation (ICRA)* (pp. 14093–14100). IEEE.
10. Hu, Y., Yang, J., Chen, L., et al. (2023). Planning-oriented autonomous driving. In *Proceedings of the IEEE/CVF Conference on Computer Vision and Pattern Recognition* (pp. 17853–17862).
11. Vaswani, A., Shazeer, N., Parmar, N., et al. (2017). Attention is all you need. In *Proceedings of the Advances in Neural Information Processing Systems*.

MIX
Papier aus verantwortungsvollen Quellen
Paper from responsible sources
FSC® C105338

If you have any concerns about our products,
you can contact us on
ProductSafety@springernature.com

In case Publisher is established outside the EU,
the EU authorized representative is:
Springer Nature Customer Service Center GmbH
Europaplatz 3, 69115 Heidelberg, Germany

Printed by Libri Plureos GmbH
in Hamburg, Germany